Adaptive Survey Design

Chapman & Hall/CRC
Statistics in the Social and Behavioral Sciences Series

Aims and scope

Large and complex datasets are becoming prevalent in the social and behavioral sciences and statistical methods are crucial for the analysis and interpretation of such data. This series aims to capture new developments in statistical methodology with particular relevance to applications in the social and behavioral sciences. It seeks to promote appropriate use of statistical, econometric and psychometric methods in these applied sciences by publishing a broad range of reference works, textbooks and handbooks.

The scope of the series is wide, including applications of statistical methodology in sociology, psychology, economics, education, marketing research, political science, criminology, public policy, demography, survey methodology and official statistics. The titles included in the series are designed to appeal to applied statisticians, as well as students, researchers and practitioners from the above disciplines. The inclusion of real examples and case studies is therefore essential.

Published Titles

Analyzing Spatial Models of Choice and Judgment with R
David A. Armstrong II, Ryan Bakker, Royce Carroll, Christopher Hare, Keith T. Poole, and Howard Rosenthal

Analysis of Multivariate Social Science Data, Second Edition
David J. Bartholomew, Fiona Steele, Irini Moustaki, and Jane I. Galbraith

Latent Markov Models for Longitudinal Data
Francesco Bartolucci, Alessio Farcomeni, and Fulvia Pennoni

Statistical Test Theory for the Behavioral Sciences
Dato N. M. de Gruijter and Leo J. Th. van der Kamp

Multivariable Modeling and Multivariate Analysis for the Behavioral Sciences
Brian S. Everitt

Multilevel Modeling Using Mplus
W. Holmes Finch and Jocelyn E. Bolin

Multilevel Modeling Using R
W. Holmes Finch, Jocelyn E. Bolin, and Ken Kelley

Big Data and Social Science: A Practical Guide to Methods and Tools
Ian Foster, Rayid Ghani, Ron S. Jarmin, Frauke Kreuter, and Julia Lane

Ordered Regression Models: Parallel, Partial, and Non-Parallel Alternatives
Andrew S. Fullerton and Jun Xu

Bayesian Methods: A Social and Behavioral Sciences Approach, Third Edition
Jeff Gill

Multiple Correspondence Analysis and Related Methods
Michael Greenacre and Jorg Blasius

Applied Survey Data Analysis, Second Edition
Steven G. Heeringa, Brady T. West, and Patricia A. Berglund

Informative Hypotheses: Theory and Practice for Behavioral and Social Scientists
Herbert Hoijtink

Generalized Structured Component Analysis: A Component-Based Approach to Structural Equation Modeling
Heungsun Hwang and Yoshio Takane

Bayesian Psychometric Modeling
Roy Levy and Robert J. Mislevy

Statistical Studies of Income, Poverty and Inequality in Europe: Computing and Graphics in R Using EU-SILC
Nicholas T. Longford

Foundations of Factor Analysis, Second Edition
Stanley A. Mulaik

Linear Causal Modeling with Structural Equations
Stanley A. Mulaik

Age–Period–Cohort Models: Approaches and Analyses with Aggregate Data
Robert M. O'Brien

Handbook of International Large-Scale Assessment: Background, Technical Issues, and Methods of Data Analysis
Leslie Rutkowski, Matthias von Davier, and David Rutkowski

Adaptive Survey Design
Barry Schouten, Andy Peytchev, and James Wagner

Published Titles (continued)

Generalized Linear Models for Categorical and Continuous Limited Dependent Variables
Michael Smithson and Edgar C. Merkle

Incomplete Categorical Data Design: Non-Randomized Response Techniques for Sensitive Questions in Surveys
Guo-Liang Tian and Man-Lai Tang

Handbook of Item Response Theory, Volume 1: Models
Wim J. van der Linden

Handbook of Item Response Theory, Volume 2: Statistical Tools
Wim J. van der Linden

Handbook of Item Response Theory, Volume 3: Applications
Wim J. van der Linden

Computerized Multistage Testing: Theory and Applications
Duanli Yan, Alina A. von Davier, and Charles Lewis

Adaptive Survey Design

Barry Schouten
Statistics Netherlands, Division of Process
Development, Methodology and IT,
The Hague Utrecht University, Faculty of Social
and Behavioural Sciences, Utrecht

Andy Peytchev
University of Michigan, Ann Arbor, Michigan
Joint Program in Survey Methodology, University
of Maryland, College Park, Maryland

James Wagner
University of Michigan, Ann Arbor, Michigan
Joint Program in Survey Methodology, University
of Maryland, College Park, Maryland

CRC Press
Taylor & Francis Group
Boca Raton London New York

CRC Press is an imprint of the
Taylor & Francis Group, an informa business
A CHAPMAN & HALL BOOK

CRC Press
Taylor & Francis Group
6000 Broken Sound Parkway NW, Suite 300
Boca Raton, FL 33487-2742

ISBN 13: 978-1-4987-6787-3 (hbk)

Library of Congress Cataloging-in-Publication Data

Names: Schouten, Barry, 1971- author. | Peytchev, Andy, author. | Wagner, James (James Robert), author.
Title: Adaptive survey design / Barry Schouten, Andy Peytchev, James Wagner.
Description: Boca Raton, Florida : CRC Press, [2017] | Includes bibliographical references and index.
Identifiers: LCCN 2017011280| ISBN 9781498767873 (hardback) | ISBN 9781315153964 (e-book) | ISBN 9781498767880 (adobe reader) | ISBN 9781351650014 (epub) | ISBN 9781351640497 (mobipocket)
Subjects: LCSH: Surveys--Methodology. | Sampling (Statistics) | Social sciences--Statistical methods.
Classification: LCC HA31.2 .S365 2017 | DDC 001.4/33--dc23
LC record available at https://lccn.loc.gov/2017011280

Visit the Taylor & Francis Web site at
http://www.taylorandfrancis.com

and the CRC Press Web site at
http://www.crcpress.com

Contents

Section II Preparing an Adaptive Survey Design

Section III Implementing an Adaptive Survey Design

Section IV Advanced Features of Adaptive Survey Design

Section V The Future of Adaptive Survey Design

Acknowledgments

We are grateful to management, researchers and survey coordinators at Statistics Netherlands' methodology, data collection and statistical analysis departments that have been involved in the implementation of various adaptive survey design pilots. Specifically, we like to thank Melania Calinescu, Annemieke Luiten and Joep Burger who assisted in the analysis of the Dutch Labor Force Survey that serves as an example to this handbook.

We thank those who have facilitated the Adaptive Survey Design interventions in the 2010–2013 data collection for the National Intimate Partner and Sexual Violence Survey (NISVS), conducted for the Centers for Disease Control and Prevention (CDC) by RTI International. The CDC team was led by Michele Lynberg Black, Mikel Walters, and Marcie-jo Kresnow. The RTI team was led by Lisa Carley-Baxter. We are especially thankful for the help from Jamie Ridenhour, Lilia Filippenko, Jessica Williams, Dave Roe and T.J. Nesius for the implementation of the interventions. The findings and conclusions in this book are those of the authors and do not necessarily represent the views of the CDC.

Finally, we are grateful to all the members of the National Survey of Family Growth (NSFG) 2011–2018 team at the National Center for Health Statistics (NCHS) and the University of Michigan (UM). The UM team is led by Mick Couper and William Axinn. The NCHS team is led by Joyce Abma and Anjani Chandra. The NSFG is conducted by the Centers for Disease Control and Prevention's (CDC's) National Center for Health Statistics, under contract # 200-2010-33976 with University of Michigan's Institute for Social Research with funding from several agencies of the U.S. Department of Health and Human Services, including CDC/NCHS, the National Institute of Child Health and Human Development (NICHD), the Office of Population Affairs (OPA), and others listed on the NSFG webpage (see http://www.cdc.gov/nchs/nsfg/). The views expressed here do not represent those of NCHS or the other funding agencies.

Authors

Barry Schouten earned his master's and PhD in mathematics, started his career as a junior methodologist at the Methodology Department of Statistics Netherlands in 2002. In 2009, he became a senior methodologist and coordinator for research in primary data collection. His research interests gradually widened from nonresponse reduction and adjustment to multimode surveys, measurement error, and adaptive survey design. In 2017, he became a professor at Utrecht University, holding a special chair on mixed-mode survey designs. He is one of the coordinators of a joint data collection innovation network (WIN in Dutch) between Statistics Netherlands and Utrecht University that was established in 2016.

Andy Peytchev is a research assistant professor at the University of Michigan's Survey Research Center, where he leads grant-funded methodological research, works on large-scale data collection and teaches in the Survey Methodology Program. Prior to that, he was senior survey methodologist at RTI International, where he was involved in the design and implementation of government-funded surveys, as well as in staff training on responsive and adaptive survey design. He earned his master's in survey research and methodology from the University of Nebraska-Lincoln and a PhD in survey methodology from the University of Michigan.

James Wagner is a research associate professor at the University of Michigan's Survey Research Center. He has been working on surveys for more than 25 years. He completed his PhD in survey methodology at the University of Michigan. He currently teaches courses on survey methods in the Michigan Program for Survey Methodology and the Joint Program in Survey Methodology at the University of Maryland. His research is in the area of nonresponse, including using indicators for the risk of nonresponse bias to guide data collection, defining decision rules about when to stop data collection, and methods for implementing mixed-mode surveys.

Section I

Introduction to Adaptive Survey Design

1

Introduction

1.1 Why a Book?

Interest in adaptive survey design (ASD) in the past several years has been apparent. Survey methodology conferences have sessions on advances in ASD, in addition to a bi-annual workshop on ASD. Progress has been further spurred on by adoption of ASD by several survey organizations. To date there are, however, very few overview papers and no monograph or edited book. With this book, we endeavor to fill that gap, and seek to gather current experiences and findings from a scattered and still emerging field of research and design.

The idea to adapt data collection strategies, or, more generally, survey design features to known or observed characteristics of sample units seems natural. Persons, households, and businesses are diverse and have different preferences for how and under what conditions they provide responses. Such adaptation has a very long history; perhaps going back as far as the early days of surveys themselves. Adaptation to different population subgroups has long ago reached interviewer training and interviewer best practices. This more implicit adaptation is usually not formalized or structured beyond the level of individual sample units. But more explicit adaptation also has historical antecedents. An early published example refers to probability sampling with quotas (Sudman, 1966), and essentially defines stopping rules or thresholds to determine whether to continue effort. However, despite the apparent logic behind adaptation, there has been a strong tendency to standardize data collection. Reasons for this are both the ease and robustness in the design and implementation of survey data collection with a standardized protocol, but perhaps also because of the traditional distance between survey managers and survey statisticians. Data from external sources (sampling frame data, administrative data or data collection process data) have been widely used in sampling design and estimation, but only occasionally in data collection itself. With the computerization of surveys, including telephone and in-persons surveys, it has become possible to make use of these survey process data, or paradata, during data collection. Further, other factors, including declining response rates and increasing costs, have made it more necessary to do so. There is a strong tradeoff between cost and quality in contemporary design features. Combining low-cost, high-error methods with high-cost, low-error methods is a natural response to the increasing pressures on surveys. For example, different survey modes have very different cost and error structures. In particular, Web surveys are relatively low cost but tend to have low response rates while in-person surveys tend to have high cost and high response. In our view, survey managers and survey statisticians should work together more closely to optimally employ the variety of survey design features and external data in order to combat declining participation. With this book, we strive to contribute to this process by showing both the practical and statistical sides of ASD.

ASD is an emerging field and we surely cannot give a detailed roadmap on how to implement such designs in every detail. The specific setting, for example, survey objectives and constraints, and the utility of available external data, are very important. Furthermore, ASDs may touch on almost any aspect of survey design and it would be a vast overstatement to state that we know how to account for all possible interactions of design choices. Even if we could enumerate many of these potential interactions, they would surely change over time anyway. By delineating open questions in this book, we hope to stimulate further research and empirical validation.

1.2 Intended Audience and Assumed Prior Knowledge

This book is aimed at both survey managers and survey statisticians. This is because we feel strongly that the expertise of both is needed in survey design, and this is especially true in the case of ASD. Survey statisticians may be very good at deriving designs that employ external data such that a certain set of objectives is optimized, but only theoretically and under conditions that may not be realistic. Survey managers are very aware of trends in survey data collection and may have a strong sense of the design features that would work under specific conditions and constraints, but may not have the time or knowledge to optimize choices under these conditions and constraints. To make the book useful to both survey manages and survey statisticians, we have included both practical and statistical content. By using a number of running examples, we hope to bridge the gap between the two types of chapters. Some of the chapters have a more statistical focus and have been moved to a separate section of the book for discussion of the advanced features of ASD and a there is a chapter in the last section of the book that outlines a method for optimizing ASDs for the control of multiple error sources. In some chapters, there are specific sections of the chapter that contain more statistical material. These sections have been delineated to aid the reader. Each chapter, and the book as whole, has been designed so that skipping the statistical content should not cause difficulties for the reader.

The book is largely a stand-alone book and requires little prior knowledge or review of the existing literature. Important references are cited but we try to also explain theory and findings without the need to read these references in detail. However, we do assume that the reader has a basic to advanced understanding and knowledge of how surveys are designed, conducted, processed, and analyzed. We will use terminology here and there that is common in survey design. In the statistical chapters, more background may be needed in places, although we try to make these chapters self-contained as well.

1.3 Outline of the Book

The book is organized into five sections. Sections I through III are practical and more focused on survey operations; they may be of interest to readers who are interested in implementing ASDs. Section IV is more statistical and may be of interest to readers who want to know more about why ASDs may be effective, even after nonresponse adjustment. In Section V we look ahead and go beyond nonresponse error to examine other sources for survey error and a chapter on open questions faced by ASDs. The sections are as follows:

- Section I: Introduction to adaptive survey design
- Section II: Preparing an adaptive survey design
- Section III: Implementing an adaptive survey design
- Section IV: Advanced features of adaptive survey design
- Section V: The future of adaptive survey design

In Section I (Chapter 2), we define and describe ASD. Given that this is an evolving field, the definitions and usage of these terms may not completely coincide with all the recent literature (Groves and Heeringa, 2006; Wagner, 2008; Lynn, 2017; Schouten and Shlomo, 2017), but we will give our definitions and be consistent with those throughout the book. ASD is often compared to responsive survey design (RSD). In this chapter, we also discuss similarities and differences between ASD and RSD. Chapter 2 describes the aims and objectives of ASD, which are crucial to their construction and implementation. Furthermore, it introduces the elements of ASD that are elaborated and discussed in detail in the subsequent chapters.

In Section II, we turn to the preparatory work that is necessary in order to make initial design decisions for an ASD. The crucial design decisions in ASD are (1) the identification of population subgroups from auxiliary data that are relevant to the survey outcome variables, (2) an inventory of the survey design features that can be used to adapt or intervene, and (3) the choice (and estimation) of indicators to assess the impact of nonresponse.

In Chapter 3, we discuss how the sample will be divided into subgroups, called strata. Each stratum can potentially receive a different set of design features. For example, some strata might be assigned to the Web mode while other strata are assigned to the telephone. This stratification is an important step, but it is not made independently of other design decisions. One feature of the strata for ASD is that they should be based on subgroups that are relatively homogenous with respect to how well they respond to different available design options. A second feature of the strata is that they should have a similar impact on estimates for survey outcome variables. We discuss the various sources of auxiliary data that are available to survey designers. The actual availability strongly depends on the setting in which the survey is conducted. We also discuss techniques to stratify the population.

In Chapter 4, we address the question of how to choose design options and potential interventions for an ASD. This chapter is not meant to catalog all possible design features. Rather, it explores aspects of design features that are relevant for ASD. These aspects include the sequence and dosage in which these features are deployed, the impact these features have on error (typically nonresponse error), and costs. The choice of design options and interventions naturally follow that of the population strata. Ideally, the options should match the characteristics of the strata, that is, they are selected based on their proven efficacy for the selected strata.

In Chapter 5, we describe proxy measures for the risk of nonresponse error and how these may be used by an ASD. To this point in time, most ASDs have been focused on controlling nonresponse error. Hence, this is our focus. Other errors can be addressed by ASD, but we leave this topic to Section V. Our focus is on nonresponse bias. Although nonresponse bias cannot usually be directly measured, we describe proxy indicators for this bias. These indicators include measures of how far apart the sample and respondents are in terms of variables from the sampling frame or other sources. Another approach is to use estimates from the current design as a benchmark, and consider whether a reduction of effort or cost leads to departures from this benchmark. We assume that design options and interventions are matched to population strata to improve efficacy of effort, that is, to reduce nonresponse bias relative to 100% response.

In Section III, we focus on the implementation and subsequent refinement of an ASD. In Section II, we discussed the three initial design decisions that need to be made. To proceed to implementation, two further steps are needed (1) the identification of operational and cost constraints and (2) the optimization of the survey design given the quality criteria and the operational and cost constraints. These two steps are not straightforward in practice.

In Chapter 6, we examine operational issues associated with the implementation of an ASD. In this chapter, we also examine the related issue of costs, and discuss the use of cost models. Operational issues concern the flexibility of survey administration systems, monitoring tools, and dashboards. Specialized systems and tools have been developed, but ASD remains a challenge as it breaks with survey tradition and is more prone to unanticipated error. Complicating matters is the fact that these types of operational constraints may be outside the expertise or scope of the survey statisticians or survey designers who were involved in initial design decisions. The cost issues are important for the optimization of an ASD. In many cases, cost parameters (e.g., the cost of a noncontact attempt in a face-to-face survey) are not known and can be difficult to estimate without simplifying assumptions. We explore methods for estimating these parameters.

In Chapter 7, we build upon these results by demonstrating how cost models and indicators of nonresponse error can be used to "optimize" an ASD. The goal of the optimization is to minimize the risk of nonresponse error by deploying design features differentially across the strata. For example, the goal of an optimization might be to allocate modes (e.g., Web, mail, and telephone) to strata with estimated response rates under each of these mode assignments such that an indicator of balanced response (e.g., the R-Indicator) is maximized for a fixed budget. To date, the ASD literature shows a plethora of options to optimize design, ranging from trial-and-error approaches, to case prioritization rules, to optimization problems that are formulated and solved mathematically. The mathematically rigorous methods for identifying an optimal design are presented in the technical sections of the chapter. In the chapter, we discuss the strengths and weaknesses of the different optimization strategies.

In Chapter 8, we examine the impact that errors in the estimated design and cost parameters have on the results of an optimization. We consider both the performance of ASD in terms of quality and costs and the structure of the ASD strategy allocation. The latter is important from an operational point of view since a stable ASD is easier to implement and less prone to error than a design that changes its form in each wave. We explore issues, such as if the estimated response rate for some stratum under a specific mode is incorrectly estimated, does the design perform about as well as it would under the correct estimate of that parameter? Also does bias or imprecision in estimated response rates lead to very different optimal design options or interventions? A goal is to identify ASDs that are relatively robust to these kinds of specification errors.

This takes us to the point of having implemented an ASD. In Section IV, we then focus on advanced elements of an ASD, some of which involve statistical theory and sampling theory. An often posed criticism to ASD is that nonresponse adjustment, after data collection is completed, may be equally effective when using the same auxiliary data. ASD may be viewed as adjustment by design. Given that ASD is operationally more demanding, the criticism has a legitimate basis. Furthermore, scholars have questioned whether adaptation may negatively affect survey precision, that is, standard errors of survey estimates. In Section IV, we address these issues.

In Chapter 9, we examine why controlling proxy indicators of nonresponse bias might be expected to actually lead to reductions in nonresponse bias. As we noted, it might be argued that rather than using ASD, it could be as effective to use the available sampling frame and administrative data in postsurvey adjustments. We have already cited empirical evidence that this is not the case. In this chapter, we also explore theoretical arguments that justify altering data collection using ASD rather than only doing postsurvey adjustments. Furthermore, we link the various proxy measures for nonresponse bias that are discussed in the literature to these more formal arguments and bias derivations. We show that many of the measures are similar and, as a consequence, provide similar signals of nonresponse bias.

In Chapter 10, we take this a step further and explore why ASD and nonresponse adjustment may be complementary, rather than overlapping activities. That is, both of these steps may lead to reductions in nonresponse bias. We discuss how adaptation to paradata and partly random allocation of design options and interventions to population affect sampling variation and accuracy. In this way, we combine consideration of bias and imprecision, which is important when any bias reduction is not to be counteracted by a loss in precision.

Finally, in Section V we open some important issues for the future of ASD. These issues concern the empirical utility of ASD, operational and implementation features of ASD, and open methodological questions to further enhance ASD. One methodological extension for which early research is available, is the extension of ASD to measurement error.

Chapter 11 looks at expanding ASD to include multiple sources of error. In this chapter, ASD to control measurement error—in addition to nonresponse error—is explored. Given the push to move surveys to cheaper modes, the ability to control measurement error may be an important criterion for future survey designs—perhaps more important than nonresponse error. More generally, this chapter develops a framework for optimizing multiple sources of error. This framework uses constraints on quality indicators for some sources of errors while optimizing other quality indicators. An important distinction that is made is between single purpose surveys, that is, with one or a few key survey outcome variables, and multi-purpose surveys, that is, that have a range of survey outcome variables. In ASD considerations for measurement error, a measurement benchmark choice can be very influential, which is demonstrated.

In Chapter 12, we end with a discussion of a research agenda for the future development of ASD. There are many open questions. In addition, given a rapidly evolving survey context, there is a need for ongoing research on the impact of the various survey design features. We bring together the various questions posed throughout the book to define a broad research agenda for ASD.

2

Adaptive Survey Design: What Is It?

2.1 Introduction

In this chapter, we introduce the concept of ASD, which, contrary to nonadaptive or uniform survey design, allows for adaptation of survey design features before or during data collection to strata that are identified based on information that is auxiliary to the survey. To introduce this concept we first need to step back for a brief overview of why we may need ASD.

We begin this chapter, in the first section, with an introduction to survey costs, survey errors, and statistical concepts such as bias and variance, as key factors motivating and informing ASDs. Some of these topics are covered in greater depth in later chapters, but are needed here to introduce ASD for readers who are unfamiliar with them as they are critical to understanding the motivation for these designs, what they entail, and their general purpose.

We then follow with a brief background on major changes in recent years that give rise to the need for these designs. We describe the origin of ASD from other disciplines and from RD in survey methodology, and provide a definition and an explanation for ASDs.

The second section offers a definition and description of ASD. It also introduces responsive design (RD), a related concept that can overlap with ASD.

Section 3 expands on the flexibility and need for customization of ASD based on study objectives. Surveys can have different objectives linked to one or more survey errors, such as cost minimization, response rate maximization, nonresponse bias reduction, or variance reduction, minimization of the risk of measurement bias, and minimization of mode-specific bias. As ASD is to help meet study objectives, it is imperative to understand that ASD needs to be customized to a particular survey. Furthermore, even the relative importance of each objective can vary across studies, leading to different designs.

Finally, section 4 in this chapter provides examples of ASD implemented on four national surveys. This is the first time we introduce these studies. We will refer to them in later chapters.

2.1.1 Survey Costs

Survey costs for a study can be fixed or they can be allowed to vary when other objectives such as number of interviews, response rates, or sampling variance are fixed. In practice, this distinction is less clear as studies often have budgeted costs and other "fixed" objectives such as number of interviews, requiring these multiple objectives to be treated as quasi-fixed. This may seem like an unnecessary distinction, but for the purpose of ASD, it is of critical importance; it determines what is being optimized and the available options. For example, a fixed cost could lead to designs that control the number of sample cases or the number of interviews; on the other hand, a fixed number of interviews with a budget that is allowed to vary could also lead to a different number of sample cases, but also to different sampling designs, different incentive structures, etc.

Groves (1989) presents the linear cost model in which cost is a function of a fixed cost component, and a cost per case that can have different values for strata/domains. We discuss this cost model and extensions in more detail in Chapter 6, but there are four limitations of the general cost model, raised by Groves that motivate both the need for ASD and the need for more complex cost models in order to implement ASD:

- *Nonlinearity.* A linear model assumes that each unit added (e.g., sampled unit, completed interview, etc.) increases costs by the same amount, yet costs per interview generally decrease as sample size increases, such as the cost of the second-stage sampling in an area-probability surveys. Costs per interview also increase while attempting to achieve higher response rates. That is, more call attempts are expended for difficult cases. This feature, if not captured, can be the main reason for experiencing costs that are higher than anticipated.
- *Discontinuity.* Many survey costs follow step functions. Sample size can be increased, for example, until a point at which additional supervisors and field directors may need to be hired.
- *Stochastic features.* There is substantial uncertainty in data collection costs, and much of this uncertainty is due to factors that are outside of researcher control. For example, the cost per interview is conditional on a particular group of interviewers; another set of interviewers may result in different data collection efficiency. Even external factors such as sporting events like the Super Bowl could affect cost in telephone surveys. We also know that for such unrelated factors, the variance in a final outcome (e.g., the number or completed interviews) is the sum of the variances for that outcome associated with each factor, contributing to substantial uncertainty in the data collection process.
- *Limited domains of applicability.* The cost structure can change with larger changes in sample size. If the survey organization's capacity

is exceeded, for example, the cost model would no longer be dependent only on the organization's own costs. Similarly, a much smaller sample size may require a different study design that would involve a different cost structure.

In addition to problems with the formulation of the cost model, there is also a limitation in how it is applied in practice. Particularly in complex data collections involving interviewers, costs can be difficult to estimate in order to build accurate predictions. For example, the cost of interviewing a person in an in-person survey is a function of the area, the characteristics of the household and selected person, the interviewer, the distance between the interviewer and the selected household, the distance between sample households that could allow the interviewer to visit multiple households on a single trip, etc. Many of these costs are confounded, reducing the accuracy of the cost prediction.

Apart from challenges in estimating costs, costs can vary in unpredictable ways. Survey practitioners suspect that elections that are associated with intensive polling and fund raising negatively affect telephone surveys. Conversely, some government surveys may experience a positive effect during census years when the government publicizes the need for every adult in the country to complete the census (in countries where the census is conducted using a survey form). Even extreme weather can affect survey costs, such as creating challenges for interviewers to reach sample households and challenges in locating displaced sample members. After a discussion of survey errors, we continue this discussion of uncertainty in the following section.

2.1.2 Survey Errors

Surveys can be used for different analytic objectives. Government surveys are typically used for official statistics such as means and proportions describing the population, but there are many other uses of survey data that include estimates of change over time, estimates of population totals, and multivariate associations. Regardless of the statistic of interest, it can be affected by error. There are several sources of survey error (for an introduction to each source of error, see Groves [1989] and Groves et al. [2009]):

- Sampling error results from collecting data from a sample rather than from the full population.
- Coverage error results from a mismatch between the target population and the sampling frame(s).
- Nonresponse error occurs when data are not collected from all selected sample members.
- Measurement errors are deviations of the responses from the true values.
- Processing error can result from coding, editing, and imputation.

Each source of error can have a systematic component as well as random variation. A systematic component is one that leads to a different survey estimate, such as a population mean, consistently across replication. For example, men tend to overreport the number of sexual partners while women tend to underreport in face-to-face interviews (Tourangeau and Smith, 1996), which are systematic measurement errors due to social desirability. Similarly, in a survey estimating academic performance and school attendance using in-school survey administration, students who skip school will be less likely to be in school and participate, leading to systematic nonresponse (Kalsbeek, Folsom, and Clemmer, 1974). There are also variable errors that do not affect point estimates such as means and proportions, often referred to as random variation. Respondents may provide inaccurate answers, but on average, they can provide the correct (or "true") value. In general, random variation can reduce the precision of the survey estimates, although for multivariate statistics such as regression coefficients they also lead to an incorrect expected value.

Statistically, this leads to the concepts of bias and variance. We will use nonresponse to illustrate these concepts. Let R denote respondents, M denote nonrespondents, and N denote the full population. The nonresponse bias in an estimate of a mean ($B(\bar{Y}_R)$) is the product of the nonresponse rate (M/N) and the difference between the mean for the respondents ($\overline{Y_R}$) and the mean for the nonrespondents ($\overline{Y_M}$). An analogous representation of nonresponse bias in the respondent mean is the ratio of the covariance between the survey variable and the response propensity ($cov(Y,\rho)$), and the mean response propensity ($\bar{\rho}$):

$$B(\bar{Y}_R) = \frac{M}{N}(\overline{Y_R} - \overline{Y_M}) \cong \frac{cov(Y,\rho)}{\bar{\rho}}.$$

Note that here the response propensity—the probability of each member of the population to respond to the survey—is assumed to be known. This is equivalent to the expected value (across all possible samples) of the product of the nonresponse rate and the difference between the respondent and nonrespondent means achieved under a given survey design. In practice, nonresponse bias is estimated in a single survey administration, so that the estimated bias is:

$$B(\bar{y}_r) = \frac{m}{n}(\bar{y}_r - \bar{y}_m),$$

where r denotes sample respondents, m denotes sample nonrespondents, and n denotes the full sample. In designs where $\bar{y}_m$ is not available, but the response propensity is estimated, often through logistic regression models, conditioning on available auxiliary information, the estimated bias is

$$B(\bar{y}_r) = \frac{cov(y, \hat{\rho})}{\hat{\bar{\rho}}}.$$

When the aim is to reduce nonresponse bias, there are three possible approaches. One is to maximize the overall response rate. When $\bar{\rho} = 1$, that is, a response rate of 100%, nonresponse bias is eliminated. That is unrealistic, and studies have shown that merely increasing $\bar{\rho}$ may not reduce nonresponse bias (Curtin, Presser, and Singer, 2000; Keeter et al., 2000). A more plausible approach is to minimize the variance of ρ, as when bias can be eliminated when it is reduced to zero. An adaptive design, for example, could target cases with low propensities in order to reduce their variability and, in turn, reduce nonresponse bias. This seems like a potentially useful approach, but a study that attempted it by identifying low propensity cases and providing higher incentives to interviewers for those cases, had no success (Peytchev et al., 2010). The authors, however, concluded that the third approach likely holds the most promise—to directly intervene on the covariance between the survey variable and the response propensity, that is, some approach that is informed by both y and ρ. In this book, substantial attention is devoted to the use of ASD to limit the covariance between the survey variable and the response propensity, to reduce nonresponse bias, and on indicators of nonresponse bias that are based on this covariance.

2.1.3 Other Challenges

In addition to survey errors, there are a number of challenges facing surveys that have spurred the development of ASDs. Especially important among these challenges are rising costs, declining budgets, and declining response rates.

Increasing Data Collection Costs and Declining Budgets for Surveys. Costs for survey data collection have been increasing, contributing to the need for more efficient survey designs. Figure 2.1 shows the cost of conducting the largest survey in the United States, the Decennial Census, between 1970 and 2010 (adjusted for inflation). The cost rose by 513% in those four decades—yet the size of the population being counted rose by only 52%. Certainly, difficulties in obtaining response to the census are a partial explanation of the rising costs. Many, if not all, surveys face similar challenges.

While costs for data collection have been increasing, budgets for some statistical agencies and surveys have been shrinking in the most recent years. For example, as the cost of conducting the U.S. Decennial Census rose so drastically in the past four decades, the budget for the upcoming census in 2020 is limited to the cost of the 2010 Census, not even accounting for inflation. Some surveys have been forced to be stopped, while others have had to be redesigned to reduce cost—such as the Survey of

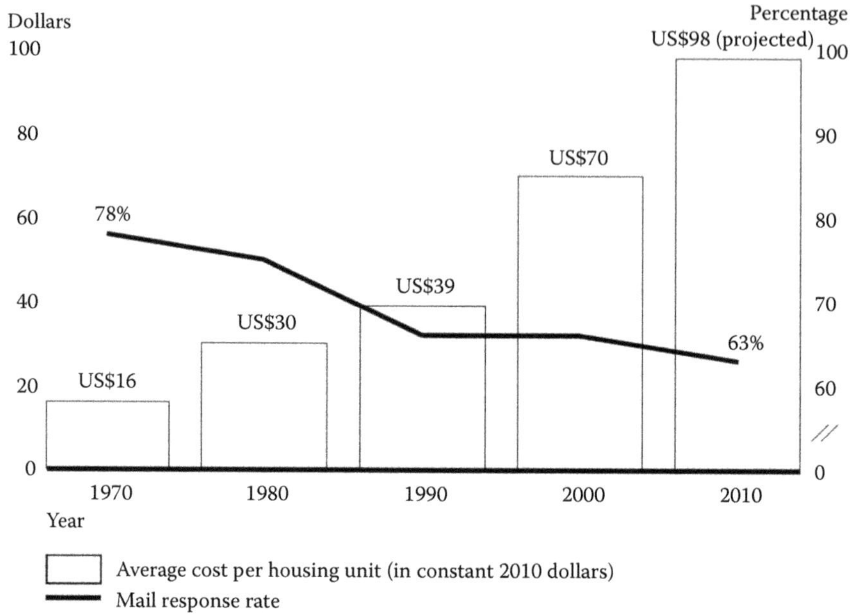

FIGURE 2.1
Average cost per household for completing the census questionnaire (in constant 2010 dollars) and census mail response rates GAO, 2013. (Adapted from GAO analysis of census bureau data.)

Income and Program Participation in the U.S., the Survey of Labour and Income Dynamics in Canada, and the Dutch Labour Force Survey in the Netherlands.

Declining Response Rates. The increasing nonresponse over the past few decades has posed multiple threats to surveys. Survey organizations have drastically increased data collection effort in an attempt to counter the increase in nonresponse (e.g., Curtin, Singer, and Presser, 2005) increasing data collection costs. Yet, as seen in Figure 2.1, despite the increased cost of conducting the U.S. Decennial Census, the response rate in that same period declined from 78% to 63%. Figure 2.2 shows the increase in the nonresponse rate in the U.S. National Health Interview Survey, one of the largest ongoing interviewer-administered surveys in the U.S. Over two and a half decades, between 1990 and 2015, the unweighted household nonresponse rate increased from 4% to 30%. Most of the increase in nonresponse occurred in the second part of this 25-year period, raising further concern. It may be even more concerning that the largest component of nonresponse that is contributing to this trend is refusal; not the failure to reach someone, which could be more random, but sample members who refuse to participate in this survey whose decision may be informed by the topic of the survey, the sponsor, etc. The need for developing approaches to address the threat of nonresponse bias in survey estimates has never been greater.

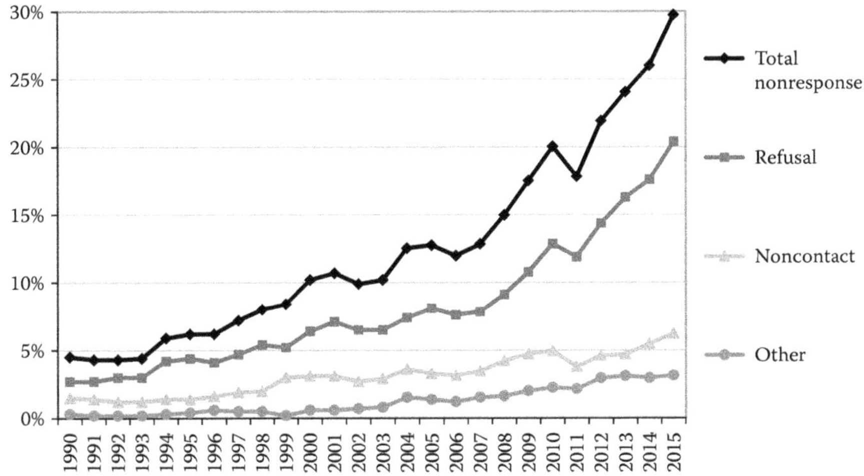

FIGURE 2.2
Unweighted household nonresponse in the national health interview survey between 1990 and 2015, by type of nonresponse. (Adapted from Stussman, B., J. Dahlhamer, and C. Simile. *The Effect of Interviewer Strategies on Contact and Cooperation Rates in The National Health Interview Survey*. Washington, DC: Federal Committee on Statistical Methodology, 2005; we are grateful to James Dahlhamer for providing an update through 2015.)

2.1.4 Need for More Flexible Survey Designs to Address Uncertainty in Data Collection

Decades of research on survey methods inform current survey designs. Many surveys have similar designs and can benefit from each other. Repeated cross-sectional surveys and panels surveys further benefit from testing and improvement over repeated administrations of their specific designs. However, there are reasons why even then, there is a need for ASD. First, there is an uncertainty in how sample members respond to different design features. One way to view this source of uncertainty is in the form of *sample heterogeneity*. An example is with gaining cooperation, some people may be more likely to respond to altruistic reasons for participation in the survey, while others may respond to a particular incentive amount. Leverage-Salience theory, first proposed by Groves, Singer, and Corning (2000) provides theoretical and empirical support for this heterogeneity.

According to Leverage-Salience theory, different survey attributes such as the topic of the survey and the use of incentives have different leverages on a person's decision to participate. For one person, the topic of the survey may have a strong positive leverage, while for another it may not have a have a strong leverage in their decision to participate—and may even have a negative leverage. Salience refers to how prominent this feature is made to the sample member. For someone for whom the topic of the survey has a strong positive leverage on their decision to participate, it would be essential to bring up the topic in contact materials, survey introductions, and other communication, in order to make it more salient relative to other features of the survey design. Yet, for the second person in this example, for whom the topic did not have a strong leverage or even had a negative leverage, exactly the opposite would be true—the topic of the survey has to be made less salient, and other features that have greater positive leverages (such as incentives) can be made more salient. This heterogeneity forms the basis for ASDs.

Second, researchers seldom know *what set of procedures will be most effective for a survey,* prior to conducting the study. The choice of mode of data collection, contact protocol, incentive type, amount and timing, are examples of major types of design decisions. Other content of the interviewer training and the text in the informed consent scripts are examples of more specific design choices. Taken together, researchers have a near infinite number of possible design combinations. The choice of design features can be informed by other surveys and reported experiments, but each study has unique aspects that limit the utility of past findings, leaving a degree of uncertainty about the optimal set of design features for a study. It may be near impossible to know whether to use a US$10 prepaid incentive at the onset of a study, US$15 prepaid incentive introduced at some point during the data collection period, US$1 prepaid incentive at the onset with a US$25 promised incentive offered late in the data collection period, or a US$50 promised incentive used in a nonresponse follow-up phase, and these are just a limited set of possibilities based on a single design feature. Indeed, while many studies find that prepaid incentives yield higher response rates, some studies do not find the same result. This finding is not surprising as survey design features interact with each other, and their effect is also conditional on what has occurred on the survey previously; the sequence of treatments. The conditional offer of an incentive can interact with the amount of the incentive, and both can have different effects on increasing participation depending on the mode and target population.

The paradigm of measuring the impact of a treatment is exactly what randomized controlled trials (RCTs) are designed to do. In an RCT, for example, subjects are randomly assigned to a treatment, such as a new drug. However, there are several challenges in applying the RCT framework to surveys, which limits the applicability to other surveys. RCTs typically involve a single treatment, while survey protocols can be seen as unique packages of treatments (e.g., prenotification letters, model of data collection, interviewer training protocols, and topic of the survey), the effects of which are unlikely to be independent. This may explain many contradictory findings in the survey literature about single survey design features in isolation from their context.

Third, there is also *natural variability* that is often beyond researcher control: variability in the environment (e.g., the impact on telephone survey data collection outcomes during periods of intensive pre-election polling or unusually cold winters for in-person surveys), variability in administering the protocol (e.g., different project staff, supervisors, and interviewers who may vary in their skills), and variability in sample composition (i.e., selecting a sample that may happen to be less cooperative relative to the rest of the sampling frame). These sources of variability are challenging to predict with any degree of certainty.

ASD has a role in addressing all three of these sources of uncertainty. Consider a survey in which the sample is heterogeneous with respect to how they respond to different protocols (e.g., mode, incentive amount), multiple protocols can be sequentially administered to sample members, and that some individuals in this sample seem to not respond to any protocol. One may want to use a survey design that builds on past data to identify what protocol works best on whom, what sequence, dosage (e.g., incentive amount) and timing of protocols to use on different groups in the sample, and maybe even use information collected during data collection to identify which sample elements are not worth pursuing further. But first, we step back to describe the common survey design paradigm and the developments that facilitate alternative approaches to survey design, before introducing ASD.

2.1.5 Common Survey Design Paradigm

Survey design is approached in various ways, but typically follows several steps:

- Identify a general design to meet the study objectives (e.g., number of interviews, level of precision and timeframe), for a given target population, within cost constraints. Make major decisions such as sampling frames, modes of contact, and modes of interview.
- Develop the detailed study design, such as training protocols, contact materials, contact attempt rules and limits, and incentive timing and amount. Make decisions among possible choices based on prior experience, research literature, and subjective judgment.
- Develop the sampling design, to include assumed response rates and number of sample cases to be released.
- Collect the data. If any assumptions are found to be inaccurate and the objectives are not being met, develop remedies. An example is increasing the number of contact attempts and changing their scheduling, when a larger than expected proportion of the sample could not be contacted.
- Document any unplanned changes to data collection. If warranted, incorporate information about these changes into the weighting and estimation.

One striking aspect of this common approach to the design and conduct of surveys is that it largely ignores the first and second source of uncertainty described in Section 2.1—the unknown optimal set of design features and variability outside of researcher control. Uncertainty in data collection outcomes is seen as a nuisance factor, and when any of the assumptions that have gone into the design of the survey fail, *ad-hoc* remedies are put into action. As the best design option among several choices may be unknown prior to data collection in the absence of experimental designs, decisions are made without direct evidence, guided only by existing literature and past experience. And regardless of the selected design, it may not perform as expected for any number or reasons, including the fact that the implementation of ad hoc changes is often hampered by time constraints that lead to suboptimal procedures. Could a researcher include multiple design options within the same survey, to test alternatives or to simply have alternatives at the ready if needed?

A second aspect of major importance is the treatment of the sample as a homogenous group. By developing a single protocol (albeit with multiple design features) and administering it to each sample member, an implicit important assumption is that this protocol is more effective than alternative protocols, for the vast majority of the target population. Numerous studies have demonstrated this third source of uncertainty in survey data collection outcomes, such as identifying sample members who complete surveys for altruistic reasons and others who are swayed by monetary incentives, and sample members who provide responses with less measurement error in one mode and others in a second mode. Could a researcher design different protocols to be used in a survey, to permit tailoring of procedures to the sample member?

These last questions lead us to the motivation for ASDs. These designs are also aided by new opportunities posed by developments such as paradata, new sources of auxiliary data, real-time data collection systems, and new analytic methods, briefly introduced in the next subsection.

2.1.6 New Opportunities

Paradata, first defined by Couper (1998), are data generated in the process of collecting data, such as clicks in a Web survey, timestamps, and outcome codes from each contact attempt. These data have grown in variety and ubiquity, particularly as a result of the introduction of computer-assisted interviewing, computer-administered survey modes, and computerized sample management systems. These data can be useful for many purposes. For example, computerized data collection systems can capture each keystroke or selection, and timing data associated with each input. Changing responses and spending more time on questions have been found to be associated with survey responses in some studies (Heerwegh, 2003) and can be used to provide insight into how to improve survey instruments. The types of paradata are described in more detail in Chapter 3 (Stratification).

Auxiliary data are also increasing in variety and availability. For example, commercial data vendors can match household-level and person-level data to sample telephone numbers and sample addresses, which in turn to could be used to improve sampling designs and data collection procedures. These commercial auxiliary data often suffer from high levels of missingness, linkage error and measurement error, yet they could prove to be useful depending on the use—such as tailoring data collection procedures based on likely household composition and characteristics.

Real-time (and near-real-time, such as daily updating) data collection systems offer an unprecedented ability to modify data collection procedures in response to outcomes. For example, simultaneous tracking of daily interviewer hours, contact attempts, refusals and number of completed interviews can help inform when to stop a particular phase of data collection and switch to a different set of procedures. In addition to yielding more timely data, real-time data collection systems also provide the ability to change procedures faster, such as switching to another mode of data collection on the following day. Even more crucial to ASDs, these systems often allow for switching procedures only for some sample cases; for example, if a sample member refuses a telephone interview, that case may be automatically assigned to another mode via rule-based algorithms implemented through computerized sample management systems.

There have also been substantial analytic developments of relevance to responsive and ASDs, in recent years. For example, Chapters 5 and 9 introduce indicators of nonresponse bias that could inform responsive and ASDs. Two key indicators are the R-indicator and the fraction of missing information (FMI). Both are model intensive, with the R-indicator relying on the estimation of response propensities and the FMI involving multiple imputation for the full sample. The combination of available paradata, auxiliary information, real-time data collection systems, and the ability to fit complex models on an ongoing basis during data collection provides a wide array of opportunities to inform changes during data collection.

There have also been theoretical developments that motivate ASD. Leverage-salience theory introduces the concept that the most effective protocol may be different across sample members. More recently, several studies have presented further experimental evidence—that some people may be enticed by the survey topic, while others by the incentive, and the two may be fairly distinct groups (Groves, Presser, and Dipko, 2004; Groves et al., 2006)—supporting the need for different treatment of sample members.

2.2 ASD and RD

Although primarily for simplicity of presentation the terms, "Adaptive Survey Design" and "Responsive Design" have been used interchangeably (Couper and Wagner, 2011), they do have different origins and, as a result, different meaning. We first present background and description of ASD, followed by RD. We end this section with a discussion on how the two types of design relate to each other.

2.2.1 Adaptive Survey Design

ASD (Wagner, 2008) is based on the premise that samples are heterogeneous, and the optimal survey protocol may not be the same for each individual. A particular survey design feature such as incentives may appeal to some individuals, but not to others (Groves, Singer, and Corning, 2000; Groves et al., 2006), leading to design-specific response propensity for each individual. Similarly, relative to interviewer-administration, a self-administered mode of data collection may elicit less measurement error bias for some individuals, but more measurement error bias for others. The general objective in ASD is to tailor the protocol to sample members in order to improve targeted survey outcomes. For example, one study used census aggregate data matched to sample telephone numbers to tailor the time of day in which to call each telephone number (Wagner, 2013). The design was further extended to incorporate paradata such as outcomes from prior call attempts in the assignment of when to call the number for the next attempt. In another telephone survey, the census aggregate data were used to identify sample telephone numbers with lower likelihood of responding, assigning them to more experienced interviewers until the first contact with a person (Peytchev, 2010).

ASD's origin can be linked back to adaptive interventions and dynamic adaptive treatment regimes in clinical trials and behavioral interventions (Murphy, 2003; Murphy et al., 2007). The basic premise of adaptive interventions is shared by ASDs—tailoring methods to individuals based on interim outcomes. We label these *dynamic* adaptive designs to reflect the dynamic nature of the optimization, and *static* adaptive designs when they are based solely on information available prior to the start of data collection (Schouten, Calinescu, and Luiten, 2013). For example, measures of depression can be obtained periodically, and interventions used when an onset of a depression episode is detected. Adaptive treatment regimes have been construed in this field as having five components:

1. A distal outcome and a proximal outcome (the main long-term and short-term goal).
2. A set of tailoring variables (i.e., patient information used to make treatment decisions).
3. Decision rules (e.g., if high risk of undesirable outcome, then implement the intervention).
4. Intervention options (type of intervention, such as medication, and dose).
5. Decision points (time at which treatment options should be considered, based on patient information).

In surveys, we often also want to know whether a design feature helped with respect to the intended outcomes. Surveys often involve repetition: surveys with continuous data collection involve the release of sample at regular intervals and maintaining an ongoing data collection under a given design, repeated cross-sectional surveys repeat the survey at usually specified intervals, longitudinal panel surveys administer the survey to the same sample over time and rotating panel designs are a hybrid design that involves both the release of new sample and reinterviewing sample members from prior iterations (waves). The inherent replication in much of survey research provides an opportunity to improve an ASD over time. Therefore, in ASD there are two more components that become more prominent due to the repetitive nature of many cross-sectional and panel survey designs:

6. Evaluation (involving some randomization of the treatment to provide direct estimates of the effectiveness of the entire design).
7. Revision (altering the design based on the evaluation, for the next implementation of the same survey).

Since the objectives and settings of clinical trials and health interventions present a large departure from those in surveys, let us make an effort to introduce these concepts in a survey context using an example. A distal outcome is the ultimate objective, for example, to complete an interview. A proximal outcome may be whether the respondent was cooperative on the last contact attempt. A tailoring variable is used to inform the decision to change treatments, such as the type of concerns the sample member may have raised at the doorstep. Decision rules would include the matching of information from the tailoring variables (concerns about time, not worth their effort) to interventions (a shorter version of the interview, a larger incentive). Finally, the decision points need to be defined, such as whether to apply the rules and intervene at the time of the interaction or at a given point in the data collection period.

One could argue that similar designs have been used on surveys prior to this theoretical framework, but without this overarching concept they can omit valuable components, involve subjective decisions that reduce replicability, and can miss out on extending these methods to more aspects of data collection. For example, some people may be likely to refuse to participate in a survey, but if these individuals can be identified, they may be willing to answer a few key survey questions. This seems to require some sort of adaptive intervention in order to leverage a short questionnaire to reduce nonresponse bias. In the British Crime Survey a short questionnaire was developed, which interviewers could choose to administer if they determine that the respondent would be unwilling to do the full interview, a method labeled as "Pre-Emptive Doorstep Administration of Key Survey Items (PEDAKSI)" (Lynn, 2003). This approach was tested on part of the sample to collect information from likely nonrespondents to use their data to inform nonresponse adjustments. One seeming deviation from an ASD of the PEDAKSI example is the administration of the decision rules *by the interviewer*; in an ASD, special emphasis is devoted to standardization of procedures and replicability, thus decisions are generally implemented algorithmically by a central office rather than the interviewer. The algorithms are based on the results from the analysis of relevant data.

An ASD needs to have clear objectives, measurable outcomes, the ability to administer different treatment regimes to different sample members, data-driven decision rules, and a plan to evaluate the effectiveness of the intervention. The last section of this chapter provides examples of ASD from four national surveys.

We now take one of those examples and describe each of the seven components. The example draws upon two similar random digit dial telephone surveys. RDD telephone surveys rely on inefficient sampling frames. Telephone numbers are selected from telephone number banks* that are known to be active or to have listed telephone numbers. Sampling from these banks yields samples with many numbers that should not be in the frame, such as nonworking, business, and fax numbers. Methods have been developed to identify such numbers, but they are imperfect and the "cleaned" samples still contain a large proportion of nonsample elements—numbers that are not active and tied to a household or individual. While these unwanted numbers do not pose a direct threat in terms of bias in survey estimates, they add considerable cost to data collection as interviewers spend time dialing these numbers, sometimes unable to determine whether they may reach an eligible sample member. Under a fixed budget, these numbers also contribute to lower response rates, as interviewers spend part of their time on unproductive numbers—instead of on those that may reach an eligible person. From an ASD perspective, we would like to identify which sample telephone numbers are to be dialed and which numbers to be placed on hold as unproductive. An approach with this objective was developed in the 2013 National Intimate Partner and Sexual Violence Survey (NISVS) and later also implemented in the 2015–2016 California Health Interview Survey, both dual-frame landline and cell phone RDD surveys. Details of how it was applied and results from NISVS are presented in Chapter 7, providing support for this application of an ASD. Here are the seven ASD components as they relate to this design:

1. Reduction of cost of data collection and increase of response rates are the two main outcomes of interest. By reallocating the interviewers' effort to more productive sample numbers, not only could cost efficiency be improved, but interviewers could also conduct more interviews.
2. The paradata from prior call attempts—the call outcomes from previous dials—provides the tailoring variables. Any prior

* Usually a set of 100 or 1,000 numbers defined by the first set of digits in the telephone number, such as (919) 412-41XX and (919) 412-4XXX, where the Xs are the randomly generated digits in that telephone number bank.

refusal, proportion of prior call attempts resulting in no contact, and encountering an answering machine on a prior call are variables that are used to tailor the data collection. In this example, these data are used in a response propensity model that is estimated using past data, and the estimated regression coefficients are applied to the current data collection to estimate the likelihood that a sample number would lead to a completed interview. Other candidate variables for inclusion in these models are baseline characteristics such as the sampling strata. Separate models are estimated for samples from different sampling frames (landline or cell phone).

3. Thresholds based on the estimated response propensities are the decision rules. If the estimated propensity for an interview for a particular sample number is below a threshold, such as 0.005, then that sample case can be classified as likely nonproductive.
4. Placing a telephone number on hold or allowing additional call attempts are the two intervention options for each sample case, (during Phase 1 in this ASD; all cases are eligible for Phase 2 selection, which involves nonresponse follow-up for a subsample of nonrespondents).
5. The decision points for assignment to an intervention option can be at several points during each data collection period. In practice, it has been implemented at a single point in time during each main data collection phase (Phase 1).
6. To evaluate the effectiveness of this ASD, part of the sample is randomly assigned to a control condition. Regardless of their estimated response propensity, none of the sample numbers in the control condition are placed on hold. This allows for an evaluation of the number of unproductive call attempts that are being avoided, as well as the potential for missed interviews due to stopping cases. Note that this design does not permit evaluation of potentially increased response rates—that would require a more complex evaluation design, likely treating each condition as a separate data collection effort each with a fixed number of interviewing hours.
7. The selected propensity threshold and the timing of the intervention call for revision based on prior evaluations. The first set of design parameters (threshold and timing) can be arbitrary, often conservative to avoid the possibility of missing interviews and attaining lower response rates. Based on continued evaluation of each implementation of the ASD, these design parameters can be adjusted to balance benefits with any adverse effects.

ASDs can vary greatly, as we show in the example case studies later in this chapter and throughout this book. In a repeated cross-sectional survey it can be viewed as experimentally implementing interventions in order to learn more about what design alternatives are more effective for the specific survey. In this sense, adapting can refer not only to tailoring to a particular sample member, but also tailoring of the general methods to a particular population and setting. This leads us to the introduction of another related term in the survey field, RD.

2.2.2 Responsive Design

RD (Groves and Heeringa, 2006) is based on two major limitations of designing one survey protocol (mode, contact materials, number of contact attempts, incentives, etc.):

- Uncertainty about the best protocol, particularly as there are a near-infinite number of alternative combinations of design features.
- Using multiple protocols during the course of data collection may be more effective than using a single protocol, as in the case of nonresponse, different features can entice different people into participation.

As a result, RD refers to survey design with multiple phases, with each phase implementing a different protocol based on outcomes from prior phases. Unlike in ASD, the survey protocol used in each phase is not tailored to sample members, with the exception of random assignment to a control condition in order to evaluate which design is more effective in a given phase. Different protocols may be used in a single phase, but the goal is to compare designs and identify the preferred design for use in subsequent phases. Depending on the survey and the particular features of each phase, this also means that RD may require a longer data collection period than an adaptive design, as the latter can operate within a single phase.

RD is aimed at the identification of more effective protocols and sets of protocols to meet the study objectives, primarily within a single data collection. It has four distinct components:

1. Preidentify a set of design features potentially affecting costs and errors of survey estimates.
2. Identify a set of indicators of the cost and error properties of those features and monitor those indicators in initial phases of data collection.
3. Alter the features of the survey in subsequent phases based on cost–error trade-off decision rules.
4. Combine data from the separate design phases into a single estimator.

In a study described by Groves and Heeringa (2006), it was unknown prior to conducting the survey whether it would be preferable to select one or two adults per household, when more than one adult resided there. There are numerous considerations involving response rates, bias, variance, and cost, and for most, the answer is likely survey- and estimate-specific. An experiment was embedded in an early phase of the survey using a random set of sample replicates, and households were randomly assigned to the one-adult or two-adult selection condition. Indicators for costs and errors were identified, and an evaluation of these indicators in the context of the relative importance of different objectives (e.g., variance, cost, and response rates), a decision was made to select only one adult for the remaining part of the sample.

An argument can be made that surveys conducted every month, quarter, or half-year in order to produce annual data can be viewed as multiple sample releases, with each month, quarter, or half-year representing a phase of the same survey. In that view, responsive and ASDs are harder to distinguish, and the distinction is also unnecessary. For example, steps 6 and 7 (evaluation and revision) described in Section 2.2.1 might be seen as a new RD phase in a repeated cross-sectional survey, even though the next implementation is a new data collection.

2.2.3 RD with ASD Features

The two designs are not mutually exclusive; rather, they can be complementary. For example, some surveys include multiple phases of data collection, each introducing a new set of design features, from changing incentive amounts to changing modes—a key characteristic of RD. Yet within each phase, not all sample members are subjected to the new protocol. Individuals or groups of individuals are identified based on the objectives of RD, such as those who are underrepresented among the respondents in the prior phases and can be contributing to nonresponse bias, if not interviewed. Statistical models are often employed to identify which sample members should receive a different protocol in the following phase—a key characteristic of ASD.

For example, the National Survey of Family Growth (NSFG) implemented a RD and an ASD. It started with three design phases—Phase 1 informed the design to be used in Phase 2, and Phase 3 changed the protocol in order to increase survey participation in a cost-efficient design—a clear example of a RD. Within each phase there were interventions that were tailored to sample households and individuals—such as prioritizing sample cases within the interviewer's case assignments in order to maximize the probability of obtaining an interview—an example of an ASD.

In some instances the distinction between ASD and RD can be ambiguous, particularly when designs contain features from both approaches. Furthermore, some researchers and organizations may use ASD to refer to RD and vice versa. Others also use derivative terms, such as Responsive Collection Design and Adaptive Total Design. This book is about ASD as defined in this chapter, yet it is useful to acknowledge that there is an overlap between the two approaches, and even in how the terms are used in the research literature and in practice.

2.3 Objectives of ASDs

A critical aspect of ASDs is the clear definition of the objectives, and their relative importance. Failure to do so can lead some to believe that both responsive and ASDs have the sole objective of reducing nonresponse bias. To the contrary, these designs are not constrained to deal only with nonresponse—it has simply been a substantial and continually increasing problem—and similarly, these designs are not constrained to bias. One way to view the possible types of objectives and their interplay is presented in Figure 2.3.

What should become apparent from Figure 2.3 is that:

- There are multiple sources of error.
- There is bias and variance from each error source.
- There are other quality dimensions and requirements imposed on the survey.
- Cost is a key constraint.
- There are tradeoffs among all these factors.

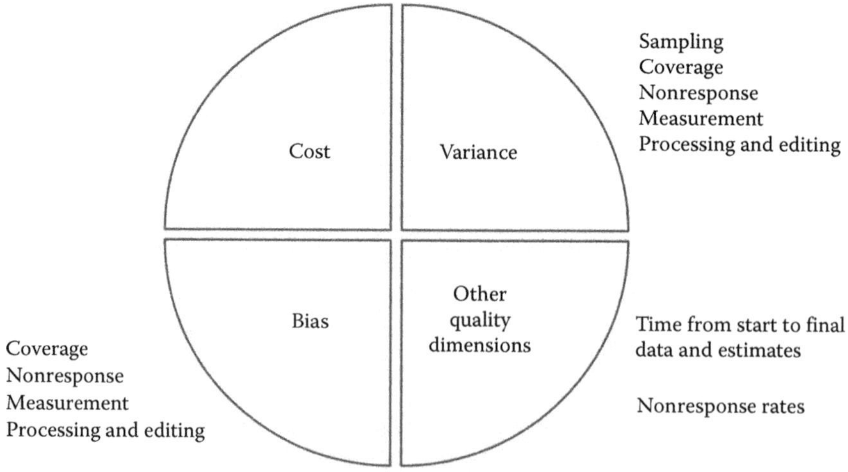

Figure 2.3 Types of survey objectives.

The last point is of great importance. ASDs aim to help with a particular objective, but very often there are tradeoffs. A key question in the evaluation of an ASD should not be its effectiveness in isolation, but whether the benefits outweigh the costs, given a certain set of priorities.

In a deceivingly simple illustration, a researcher may be interested in reducing nonresponse bias. To do so, some of the remaining nonrespondents are directed to receive a more effective and costly protocol. The objective may seem clear: reducing the risk of nonresponse bias. However, in order to meet this objective, a relatively smaller sample size might be achieved given that greater costs are incurred attempting to recruit a subset of nonrespondents.

The simplicity is certainly deceiving as the design is not specific. *How* it is specified and implemented depends on a complex evaluation of the multiple goals of the study. Figure 2.4 aims to represent the relative importance of multiple objectives on three studies: the High School Longitudinal Study (HSLS), NSFG, and NISVS.

While all three studies have the stated objective of reducing the risk of nonresponse bias, and implement multiple phases of data collection, their designs vary drastically as a function of the survey setting and the relative importance of each dimension. The HSLS places a premium on minimizing variance and less on response rates and cost, so it does not use any subsampling of nonrespondents. Nonrespondents deemed to be underrepresented among the respondent pool are identified and administered a more costly protocol involving substantially higher monetary incentives. For both NSFG and NISVS, response rates are of greater concern and minimizing variance inflation is of relatively lower priority. In these surveys, nonrespondents are subsampled.

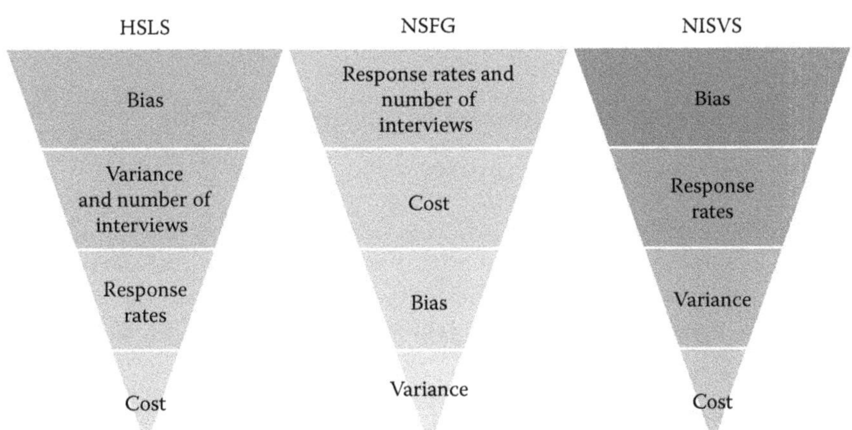

Figure 2.4 Variation in rank ordering of survey objectives in three surveys with a nonresponse follow-up phase, the High School Longitudinal Study (HSLS), the National Survey of Family Growth (NSFG), and the National Intimate Partner and Sexual Violence Survey (NISVS).

2.4 Example Case Studies

In this section we introduce several studies as examples of adaptive designs. We will also refer to these studies throughout this book and the following descriptions should give the reader a useful background on each study.

2.4.1 National Intimate Partner and Sexual Violence Surveys

Adaptive (and responsive) designs can help achieve very different survey objectives. Surveys can have multiple objectives, which need to be prioritized. Even then, multiple adaptive and RD features may need to be incorporated in order to tailor the design to meet these objectives. Multiple features may also be needed in order to balance tradeoffs between these objectives in instances when optimizing for one objective has an adverse impact on outcomes related to another objective.

NISVS is a survey of the noninstitutionalized English or Spanish speaking population in the United States aged 18 or older. It is based on a dual-frame random digit dial (RDD) telephone survey design, including both landline and cell phone numbers, collecting over 10,000 interviews each year. The survey included measures of sexual violence, stalking, intimate partner violence, and health consequences of these forms of violence. We discuss the first four years of the survey. The survey had a continuous data collection design from 2010 to 2012 with new samples released every quarter. In 2013 the data were collected during part of the year. This survey included multiple adaptive and RD features in order to meet a diverse set of objectives. We use this survey to highlight three adaptive design features, implemented at the case and subgroup levels: propensity-based assignment to experienced interviewers, propensity-based stopping of cases, and performance-based phase duration.

2.4.1.1 Propensity-Based Assignment to Interviewers

Response propensities are not a fixed person-specific characteristic. They are also a function of the design. For example, some type of people might have a higher probability of responding to a Web survey than a telephone survey. If response propensities were a fixed characteristic of persons, data collection designs would not include multiple contacts with the same sample member. Moreover, what happens on one contact attempt can affect the outcome of the next attempt; an experience with an interviewer who does not explain the purpose of the call and leaves the sample member angry can undoubtedly reduce the propensity of an interview on the next contact attempt. Conversely, it may be prudent to have a good interaction on the first contact in order to increase the likelihood of an interview, whether on the first contact or on a subsequent contact. This may be particularly important in geographic areas where people tend to be more reluctant to participate in surveys.

An experiment was conducted in which census aggregate data at the block group (statistical divisions generally containing 600 to 3,000 people) and census tract (larger areas, generally with 1,200 to 8,000 people) level were used to estimate the likelihood of participation. This model was estimated on data from one quarter and used to score the sample for the next quarterly data collection. Sample numbers with lower predicted response propensity were assigned to be dialed by more experienced and better-performing interviewers until the first contact with a sample member. This assignment led to seemingly higher interview rates among the low-propensity cases in the treatment group compared to the low-propensity control group, without any adverse effects on the high-propensity cases.

2.4.1.2 Propensity-Based Stopping of Sample Cases ("Interactive Case Management")

RDD surveys in the United States use sampling frames with a large proportion of nonsample elements—nonworking numbers, business numbers, fax and modem lines, etc. Response rates have also reached relatively low levels, despite increased calling effort (Curtin, Presser, and Singer, 2005). Combined, these factors lead to challenges in collecting data efficiently, with a large proportion of calls typically resulting in nonproductive outcomes (e.g., ringing without an answer, voicemail, and answering machine). To address this inefficiency, extensive modeling, and simulation was conducted to estimate response propensities during data collection and determine which sample numbers to stop calling, without losing potential interviews.

2.4.1.3 Phase Duration

A multi-phase design can be an effective tool to gain participation from an inherently heterogeneous population by employing multiple protocols. When these protocols also differ in cost, a multi-phase design can also be a cost-saving method by employing the more costly protocols in later phases. This was the design used on NISVS, with respondents being offered US$10 during the main data collection (Phase 1), lasting approximately 12 weeks, and a subsample of nonrespondents being offered an incentive of US$40 during the following 4–5 weeks (Phase 2).

The duration of each phase can be fixed, yet the optimal point at which to stop Phase 1 and start Phase 2 is unknown. Even if extensive simulations identify what that point in time is, it will surely vary from one quarter to another based on interviewer performance, sample performance, external factors (e.g., competing calls near election periods), time of year, etc. Thus, a key need for an adaptive design arises for the real-time identification of when to switch phases, and do so separately for the different samples.

2.4.2 The Dutch Labor Force Survey

The focus in most adaptive designs is on unit-nonresponse, either through bias, variance or costs. When the survey mode is one of the primary design features that can be adapted, then this focus is too narrow; the survey mode also impacts coverage of the population and measurement error. In redesigns of the Dutch Labor Force Survey (LFS), the choice of survey modes played a decisive role, mostly for cost reasons. Three modes (alone and combined with each other) were considered: Web, telephone, and face-to-face.

The LFS is a rotating panel survey among the Dutch population of 15 years and older. The main topics are employment, unemployment, occupation, type of economic activity, and educational level. The survey follows EU regulations and is conducted throughout the EU in a relatively similar design. The survey has monthly samples of around 12,000 addresses from a two-stage sampling design: the first stage consists of municipalities and the second stage of addresses that are stratified based on age, registered unemployment, and ethnicity. Since the addresses are sampled from municipality registers, they can be linked to government administrative data on the persons registered at the addresses.

The LFS has existed since the 1980s but went through various major redesigns. Around 2000 the survey changed from a repeated cross-sectional survey to a rotating panel survey with five waves and three month time lags. The first wave was face-to-face and the other four waves were through the telephone. In 2008, following the fast increase in population coverage of the Web and given the large costs of the face-to-face and telephone design, it was decided that the survey should use multiple survey modes in the first wave, including especially cheaper modes. Following this decision, various strategies were tried and were run parallel to the face-to-face strategy. Strategies involving the Web always had the Web as the first mode and interviewer modes as a follow-up, that is, as a second phase. The parallel runs showed that mode-specific measurement bias, or, more generally, mode effects, occur and vary in size over different population strata.

The finding of measurement bias differences led to a discussion about a benchmark mode for measurement error and about the utility of the interviewer modes in the follow-up: What mode should be considered to provide the best measurement properties for LFS topics? If the Web is considered the best measurement mode, does the improvement in nonresponse bias of the follow-up modes outweigh the loss in measurement bias? If face-to-face is the best measurement mode, does the cost reduction of the Web first phase outweigh the loss in accuracy due to measurement bias? ASDs were evaluated in the Dutch LFS on their potential to form a middle ground option. Other evaluations challenged the assumption of an optimal mode for the entire sample and involved the identification of an optimal mode for different sample members, with regard to minimizing measurement error.

2.4.3 National Survey of Family Growth

The NSFG has been conducted periodically since 1973. The survey topic focuses on fertility, marriage and cohabitation, and sexual activity. The target population is women and men ages 15–44. The survey is conducted by in-person interviews. There are two stages of interviewing. In the first stage, a "screening" interview is conducted in which household rosters are collected, including age, sex, race, and ethnicity information on all household members. The purpose of the screening interview is to identify eligible persons in the household. One eligible person is randomly selected from each household. In the second stage of interviewing, a "main" interview of 60–85 minutes is requested from the selected person. There is an ACASI section that includes questions about risky sexual behavior, drug use, sexual orientation, and nonvoluntary intercourse. The sample is selected via an area probability design with neighborhood clusters or "area segments" assigned to interviewers within primary sampling units (PSUs) which are counties or groups of counties.

The three most recent waves have included responsive or adaptive design elements. The NSFG Cycle 6 was conducted from March 2002 to March 2003. The RD elements are described in Groves and Heeringa 2006. NSFG Cycle 6 was conducted in three phases. The first phase was 9 weeks and included a quarter sample of PSUs. A US$40 incentive was offered for completing the main interview. There was no incentive for completing the screening interview. The first phase was used to determine the point at which interviewer effort (as measured by calls) ceased to produce changes in estimates. This phase led to a callback limit of 10–14 calls being placed on the sample released in later phases.

In addition to implementing the callback limits, the second phase included all PSUs. The sample was monitored on a daily basis during this phase. Response propensity models were estimated on a daily basis and used to predict the probability of response for the next call to be made. These predictions were used to direct interviewers to area segments with a relatively large sum of the propensities. The estimated propensities were also used to stratify the sample selection for Phase 3. The area segments were divided into 16 strata based on the cross-classification of quartiles of the number of active cases in each segment and the sum of the estimated response propensities in each segment. Higher sampling rates were used for segments with more active cases and larger summed propensities. The ratio of the largest to smallest sampling rates was 4:1. As Groves and Heeringa note, "... this design option placed large emphasis on the cost efficiency of the phase 3 design to produce interviews, not on the minimization of standard errors of the resulting data set" (p. 451). Several features of the third phase design were changed. The active sample was consolidated in the hands of the most productive interviewing staff. Proxy interviews were allowed for the screening process (subject to permission from a supervisor). A prepaid

US$5 incentive was mailed to unscreened cases and a US$40 prepaid incentive was sent to sampled persons selected for the main interview. This raised the incentive for the main interview to US$80.

This design was modified before the next cycle began (see Wagner et al. [2012], Kirgis and Lepkowski [2013], and Lepkowski et al. [2013] for details). The NSFG 2006–2010 adopted a continuous design. For this design, a sample of PSUs was released each year, while a new sample of area segments and housing units (within each PSU) was released each quarter. Each quarter was 12 weeks long. At the end of the 12-week period, the sample was closed out and no additional effort was allowed. NSFG 2006–2010 remained a face-to-face interview with two stages (screening and main interviewing). The new design included two phases. In the first phase, interviewers focused on completing screening interviews. The experience from Cycle 6 was that interviewers prefer conducting main interviews. This led to delays in the screening process. In NSFG 2006–2010, there was an emphasis on early screening. This included a "screener week" in week 5 that showed demonstrably higher rates of conducting screening interviews. Phase 1 also included targeted prioritization of subgroup cases whose response rates lagged behind. Phase 2 was implemented for weeks 11 and 12. A subsample of about 1/3 of the cases was selected. The incentive was raised from US$40 to US$80 for the main interview (US$40 prepaid) and a US$5 prepaid incentive was mailed to unscreened cases.

A similar design was implemented for the next iteration of the survey, NSFG 2011–2019. This design replicated many of the features of NSFG 2006–2010. New measures of interview quality were added. This dashboard used keystroke data as well as other paradata to identify interviewers who complete interviews more quickly than average, use "don't know" or "refusal" response options more frequently, and other such measures. These indicators are then used to identify interviewers for whom additional follow-up may be required. The NSFG 2011–2019 has also conducted additional experiments with new incentive amounts.

2.5 Summary

This chapter serves several purposes. It provided a background to some of the motivating factors behind ASD. Invariably, rising survey costs and declining survey participation are among them, but there are also important other factors, such as: theoretical developments supporting that the most effective survey protocol can be different across sample members, availability of more sophisticated sample management systems and variety of data sources including improved ability to harness these data in informative statistical models.

The chapter also provides a background and description of ASD, a description of RD, and the possible complementary relationship between the two approaches. We paid particular attention to the defining role of the objectives of an adaptive design, highlighting that they can be used for very different goals and that even the relative importance of one objective over another (e.g., response rate maximization vs. nonresponse bias reduction), may lead to a different design.

It also introduced three studies that employ ASD: A telephone survey and a face to face survey from the United States, and a multi-mode survey from The Netherlands. We refer to these studies in the following chapters.

Section II

Preparing an Adaptive Survey Design

3

Stratification

3.1 Introduction

One of the key decisions to be made in an adaptive design is how to divide the sample into strata, or subgroups, in order to define targeted protocols for each of the strata. We adopt the term "stratification" in order to link the term to sampling theory that defines strata as mutually exclusive and exhaustive subgroups within the population. In sampling theory, strata that are homogenous with respect to the survey outcome variable are preferred. In our usage, the term "stratification" will refer to mutually exclusive and exhaustive subgroups within the population, but useful and efficient stratification will depend upon more than just the homogeneity of the strata with respect to the outcome variable.

Stratification for ASDs should be undertaken with three objectives in mind. *First, the stratification should identify subgroups that are influential on estimates.* We will examine several ways in which this influence can be identified during data collection. The goal is to identify some strata that, when their response rate changes, have influence on the bias of final, adjusted estimates. Most adaptive designs have focused on nonresponse bias, but a focus on other sources of bias is also possible (see Chapter 11 for a discussion of adaptive design aimed at controlling measurement error). For sampling purposes, making strata as homogenous as possible with respect to the survey outcome variables is the goal. For ASDs, the goal is different. Taken to the extreme, in the case of strata that are completely homogenous—that is, all cases have the same value for the outcome variable—no adaptive design is needed. In this case, postsurvey adjustment will eliminate any nonresponse bias. However, in practice, attaining this level of homogeneity with respect to the survey outcome variables virtually never happens. For ASDs the goal is to create subgroups based on their ability to influence estimates. This latter task is more difficult to do as we do not know with certainty how any case might influence final, nonresponse-adjusted estimates until after we have collected that case. There are several practical approaches to the problem of creating strata that can influence estimates. For example, this can be done by balancing response rates across subgroups in the sample, identifying cases that are difficult to predict their outcome variable, or estimating influence via simulation. We will explore these approaches in more detail later in this chapter.

The problem of identifying strata that are influential on estimates is further complicated by the fact that most surveys have many survey outcome variables. Therefore, some cases might be best grouped with one set of cases for some survey estimates and with an entirely different set of cases for other estimates. In a worst case scenario, the stratification schemes for each estimate could be completely different. This problem is akin to the sample design problem for surveys with multiple outcomes. One solution is to identify strata that are as influential as possible with respect to a subset of "key" statistics or even a weighted composite of this subset.

The second feature is that the strata are likely to participate at different rates depending upon the protocol that is applied. Ideally each stratum would have a unique protocol that maximizes its response rate. For example, in a survey with two available modes (mail and Web), a stratum composed of persons 18–44 might achieve their maximum response rate with a request to complete a Web survey, while another stratum of persons 45+ might attain their maximum response rate with a request to complete a mailed survey. In this case, each stratum can be assigned its own protocol, rather than assigning a single protocol to all cases. Of course, the choice of protocols is subject to budget and other logistical constraints. Further, different goals might lead to the selection of protocols other than those that result in the highest response rates for each group. For instance, if maximizing a balance measure such as the R-Indicator (see Chapter 9) is a goal, and some stratum responds at a high rate for a relatively inexpensive protocol, but at a somewhat higher rate for more expensive protocols, the relatively inexpensive protocol might be chosen so that more resources could be devoted to strata that respond at lower rates under either survey design protocol.

The third feature is that some strata may be given lower cost strategies. This is implicit in the second feature. Here, we make it explicit. Since survey budgets are almost always limited, costs constrain the quality of survey estimates. Given this constraint, one objective of stratification might be to identify strata that may be given a strategy that is relatively low cost and for which the increase in the total survey error based on this strategy relative to more expensive strategies is not as great as it is for some other strata. For example, it might be the case that some strata have similar response rates whether they are offered a Web survey or recruited by telephone. If the resulting increase in nonresponse error for the Web survey relative to the telephone survey is small, then this stratum may be allocated to Web, thereby preserving budget for other strata, for whom the increase in nonresponse error when given the Web survey as the only option is higher.

This chapter begins by examining the goals of stratification more carefully in Section 3.2. Then, several methods for defining strata will be discussed in Section 3.3. Next, since an important constraint on ASDs is the availability of data with which strata can be defined, we discuss commonly available data—paradata and sampling frame data in Section 3.4. The chapter concludes with examples. The concept of strata defined in this chapter is an important "building block" for understanding ASDs. This concept will be referenced in later chapters.

3.2 Goals of Stratification

We have defined three goals for creating strata

1. Create strata that are influential on estimates.
2. Create strata that respond to different recruitment strategies in similar ways.
3. Create strata such that at least some strata may be given lower cost strategies.

In order to meet the first goal, the strata need to be defined in such a way that if that stratum is underrepresented, the estimates may be biased. This might be the situation, for example, if the strata are defined in order to be homogenous within and heterogeneous between strata with respect to the survey outcome variables. Thus, an underrepresentation of any groups indicates selectivity in the response process due to the survey outcome variables. For example, if we stratify a sample of women by age, and the survey outcome variable is the number of live births, then we might expect that an underrepresentation of older women would lead to a downward bias in estimates. Of course, it might be argued that if we simply used the age variable as an adjustment factor, we could make these biases go away. However, these adjustment models are rarely perfect since they do not meet the assumption that the model is correctly specified. In a weighting class adjustment, this assumption requires that responding and nonresponding cases within any cell have the same mean. It might be argued that using the same variables to create strata for an adaptive design as will be used in the postsurvey nonresponse adjustments is pointless, since the same bias will be eliminated by the adjustments as would be eliminated by an adaptive design. As discussed in Chapter 2, there is some empirical evidence that this is not the case. Further discussion of this issue is given in Chapter 10. It appears that ASDs can limit nonresponse bias further than just using postsurvey adjustments, even when the same variables are used in both the adaptive design and the postsurvey adjustments. This would be the case if the ASD, by controlling the observed covariates, also exerts some control on the unobserved covariates when adjustment does not lead to balance on the unobserved covariates. In Section 3.3, we will discuss several strategies for identifying strata that are influential on estimates.

If we are to identify strategies that are well-matched to the targeted cases, it may be useful for the selected groups to have a label. For example, if the cases are identified using a propensity model that is based on paradata, and low-propensity cases are targeted, it may be difficult to identify a strategy that will be more effective with these cases. The label "low propensity" does not help us identify recruitment strategies that may be effective with that group. Unless, of course, the proposed action is stopping any attempts to interview them, or the opposite action, that is, prioritizing or placing more effort on the cases that are being targeted. An approach that identifies subgroups in particular categories (i.e., groups that have labels) is articulated by Schouten and Shlomo (2015). They present an example where the targeted strata are defined using partial *R*-Indicators. In brief, partial *R*-Indicators identify variables, the categories of which have the most variation in response rates (see Chapter 9 for a more complete description). In the example given by Schouten and Shlomo, during the first phase, cases are randomly assigned to either Web or mail. After this mode has been implemented, a new response propensity model is estimated and the partial *R*-Indicators are updated. These are used to identify strata that are responding at lower rates. Strata that have lower response rates are selected for follow-up with face-to-face contact attempts.

Another goal for the creation of strata might be to create strata that can be recruited with a lower cost strategy. This may not always be necessary, but it is often necessary given the sharp budget constraints that many surveys face. In this situation, creating some strata that may be given lower cost strategies provides the survey designer a way to balance errors and costs. If some strata are given lower-cost strategies, then resources may be reallocated to other strata for which the resources will provide greater reduction in error. Members of a particular stratum, for example, older males, might respond to a mailed survey at reasonable rates and provide reasonable quality data. Of course, if that same stratum is approached with a face-to-face survey, they are more likely to respond and provide somewhat higher quality data. However, given a fixed budget, it may be better to use the mailed survey for the older male stratum and use the face-to-face strategy for those strata where the result is a greater reduction in error (both nonresponse and measurement error).

It is also worth considering robustness when defining strata. Since the strata may be based upon estimated relationships in existing data, these strata may be subject to sampling error. Creating strata that are extremely fine (i.e., having relatively few cases in each stratum) or that are based on estimated models can lead to results that are somewhat less efficient than expected. This is similar to the problem of selecting models for prediction. "Overfitting" occurs when models are selected that include predictors based on features of the data that are unique to the data in hand (i.e., due to sampling error). Methods such as cross-validation can help to avoid these issues. We consider robustness in more detail in Chapter 8.

3.3 Defining Strata

This section will discuss several methods for creating strata. As ASDs are a developing methodology, this list is not exclusive. A large class of methods is available, including expert judgment and trial-and-error. In this section, we will focus on three methods that have been used by other surveys and that can serve as a model. These three methods are partial *R*-Indicators (including some discussion of potential predictors in the models that underlie these indicators), regression diagnostics, and simulation.

It is possible to create strata that may lead to increased bias. In sampling, poorly designed strata—that is, strata that do not create homogeneity with respect to the survey outcome variable—will lead to an inefficient sample design. Design effects will be relatively larger than a design with a better stratification scheme. For adaptive designs, poorly designed strata can lead to a reduced impact of the adaptive design or may even lead to increased bias.

Given the uncertainty about the impact of the strata on nonresponse bias, it may be wise to consider several alternate stratification schemes and see how similar the results are across these schemes. If the results are very similar, then the design may be robust to the choices that can be made regarding stratification. If a survey is repeated, then the stratification may be refined over time in a trial-and-error approach. In the next section we discuss several methods for defining strata. These methods are examples of approaches that can be taken, but is certainly not an exhaustive list of all possible options. We also refer to Chapter 8 where the sensitivity of the performance of ASD to bias and sampling variation in survey design parameters, for example, contact propensities or costs per sample unit, is discussed. This sensitivity may lead to a preference for more parsimonious stratification.

The methods are response propensity variation, regression diagnostics, and simulation. Each of these methods relies upon having existing data. These data may come from an existing survey. The existing survey, ideally, would share the main features of the survey being planned. In a repeated, cross-sectional survey, prior repetitions of the survey are a good source of data. These methods differ in that some focus on the relationship between covariates from the sampling frame and paradata available for all cases and response (response propensity variation) while others focus on the relationship between these fully observed covariates and the survey outcome variables. However, as was emphasized earlier in the chapter, both these dimensions are important and need to be considered under any selected method. For instance, it may be wise to estimate partial *R*-indicators (see below and Chapter 9) only from fully observed covariates that have correlations with the survey outcome variables.

3.3.1 Response Propensity Variation

One common way to identify strata for ASDs is to use response propensity modeling to identify predictors of response. However, it is important to keep the other goals of stratification in mind. In the context of response propensity modeling, this usually means identifying predictors in the models that are also related to survey outcome variables or indicators of likely influence on estimates.

Partial *R*-Indicators may be used in adaptive designs to identify strata. Partial *R*-Indicators identify subgroups that are responding at lower (and higher) rates (see Chapter 9 for a complete description of Partial *R*-Indicators). Partial *R*-indicators decompose the variance of estimated response propensities over variables and categories within variables. These indicators are then helpful in identifying which variables explain most of the variability of response rates and, within these variables, which categories have the highest and lowest response rates. The lower-responding groups identified in this manner can then be targeted with additional or more intensive effort (e.g., a mode change).

The selection of the predictors for the model used to generate the partial *R*-Indicators is an important decision. The predictors should meet the requirements described at the beginning of this chapter. That is, they should predict the survey outcome variables and also create groups that have different response propensities under the different available designs. In some cases, the design change might be simply additional effort under the current design (e.g., more calls) that only the targeted cases receive. In this case, the design is meant to allocate scarce resources to under-represented groups.

Schouten and Shlomo (2017) discuss how to approach the identification of strata using Partial *R*-Indicators with a "structured trial-and-error approach." They suggest the following method. The survey should be divided into phases. The first phase is used to collect data about who is responding. After the first phase, a logistic regression model predicting response during the first phase is estimated. Important predictors of the survey outcome variables should be included in this model. The model estimates are then used to identify groups that are responding at lower rates. These groups might be identified as categories of a single variable, or even by the cross-classification (or interaction) of more than one variable. Then groups identified as the lowest responding groups can be treated as ASD strata (see Schouten and Shlomo for additional details).

A similar method can be applied to repeated cross-sectional surveys. The stratification can be improved across the waves by using information from prior waves to inform modeling choices. We give an example of the procedure at the end of this chapter.

An empirical demonstration of the important role played by model selection when response propensity models are used to inform adaptive designs came from Rosen et al. (2014). They used estimated response propensities to target low-responding cases. There initial models included paradata, such as number of call attempts and whether a case had ever refused. These models were estimated for the a 2012 follow-up survey with those originally interviewed in the High School Longitudinal Study of 2009 (HSLS:09). The models, which included paradata, created a wide distribution of estimated response propensities. These paradata measures were highly associated with response.

They intervened by allocating expensive, face-to-face contact attempts to the low responding cases. They were also able to increase the response rates for the lowest propensity cases in this way, while the rest of the cases received the standard telephone treatment. Although the face-to-face contacts increased the response rates for the cases with low estimated response propensities, this did not result in a change to estimates. It turned out that the underlying estimated response propensities were unrelated to the survey outcome variables.

As a result of these findings, in the 2013 wave of the same study, a different model was used to estimate response propensities. This model only included strong correlates of the survey outcome variables as predictors. The predictors were drawn from previous data collections and are, therefore, not time-varying. The predictors included enrollment status at last follow-up, race, grade when student took Algebra 1, student's final grade in Algebra 1, and National School Lunch Program status. The survey designers called this model, with only strong predictors of the survey outcome variable, the "bias likelihood" model. In order to aid interpretation, the predictions from this model were rescaled as one minus the estimated propensity. Therefore, a higher bias likelihood score indicates a type of case that has a lower response rate to this point.

Figure 3.1 shows a scatter plot of the estimates from the two models. The estimated response propensity from the paradata-driven model is on the x-axis while the estimated response propensity from the bias likelihood model with only strong correlates of the survey outcome variables is on the y-axis.

In this case, the goal was to interview the cases with the highest estimated "bias likelihood" score—that is, the score that is likely to be related to the survey outcome variables. This might be done by targeting cases that have a relatively high likelihood of responding as estimated from the paradata-driven model while having a low propensity of responding when this quantity is predicted using strong correlates of the survey outcome variables. These cases are in the upper right-hand quadrant of the figure.

These results also point to the importance of model selection when using measures based upon response propensity models. If the predictors are not related to the survey outcome variables, then the adaptive design that uses them as inputs will be less likely to reduce nonresponse bias.

3.3.2 Regression Diagnostics

Another measure of potential "influence on estimates" can be developed using survey outcome variables. Frame data and paradata are used as independent variables in regression models predicting the survey outcome

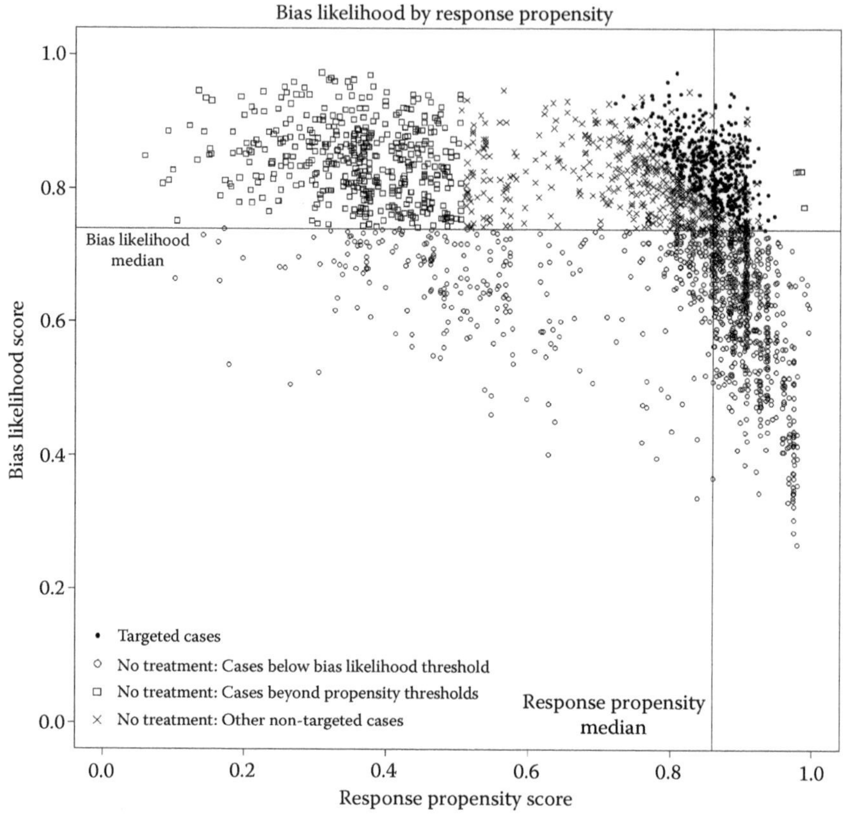

FIGURE 3.1
Bias likelihood score plotted against propensity score.

variables. The model estimates are used to identify cases that are influential, which may also be described as cases where the model provides poor fit.

Figure 3.2 demonstrates the concept of influential points in regression. The data are hypothetical, so the scale of x and y axes do not matter. The upper two figures show how a point can be influential. In these figures, the mass of data points in the lower left is the same in both figures. The large solid dot is a single, influential point. Regression lines that are fit to the data are also included in each figure. The regression line changes a great deal depending upon where this point is located. Statistics textbooks often suggest obtaining additional data in the regions where influential points are identified.

Figure 3.3 shows two different scenarios for how additional data might impact the estimated slope of the regression line shown in the left-hand part of Figure 3.2. In the figure on the left, additional data share the same y attribute as the cases that have lower values of x. The result is that the regression line has a much flatter slope than the line shown in the left side of Figure 3.2. The originally observed influential point now seems somewhat aberrant. In

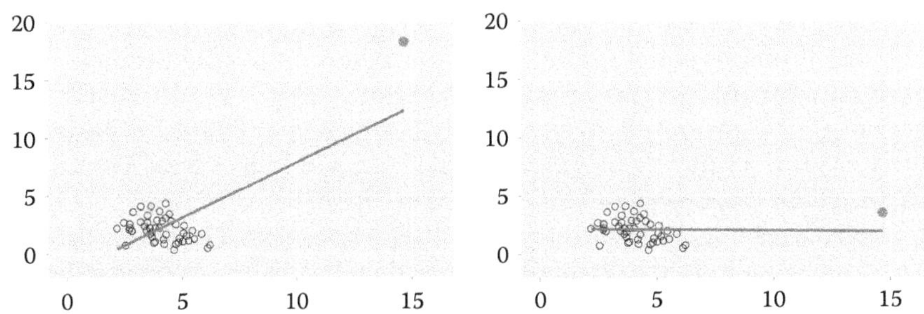

FIGURE 3.2
Examples of influential points.

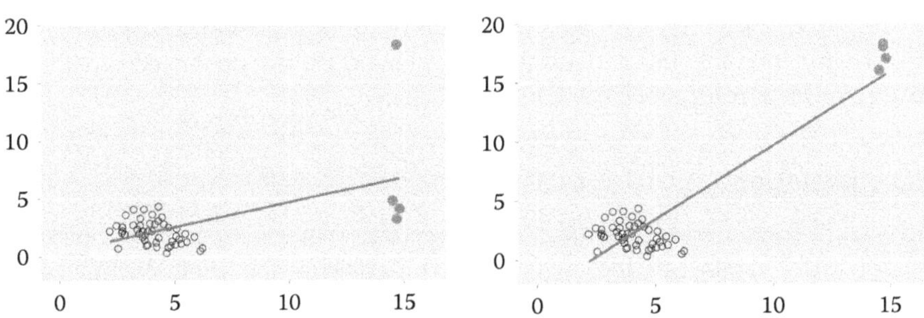

FIGURE 3.3
Addition of more data.

the figure on the lower right, the additional data confirm the original data point. These cases with a higher value of x also have a higher value of y. The estimated slope remains relatively steep.

There are several potential diagnostic measures that can be used. Many of these involve the diagonal elements of the hat matrix: $h_{ii} = x_i'(X'X)^{-1}x_i$. These are sometimes called the leverage scores. Points with higher values for h_{ii} are farther from the center of x space. For linear y_i, there are several candidate measures. These include Cook's distance

$$D_i = \frac{e_i^2}{p}\frac{h_{ii}}{1-h_{ii}} = \frac{(\hat{\mathbf{y}}_{(i)} - \hat{\mathbf{y}})'(\hat{\mathbf{y}}_{(i)} - \hat{\mathbf{y}})}{pMSE}$$

where e_i^2 is the squared residual of the ith case, p is the number of predictors in the model, and $\hat{\mathbf{y}}_{(i)}$ is the vector of predicted outcome variables when the ith case is deleted. Cook's distance might be preferred over other diagnostics since it focuses on the fitted values. Other diagnostics examine the impact on estimated coefficients. There are similar diagnostics for logistic regression models. See Belsley, Kuh, and Welsch (1980) for a full discussion of regression diagnostics. In general, cases can be influential when there are few observations in some region of values taken on by the predictor variables or when there is high variance in a region.

Wagner (2014) proposed to use regression-based influence measures to identify cases that may be useful to interview. Used in this way, a stratum of cases for which the survey outcome variables are difficult to predict is identified. These cases may be useful to include as responders. However, this stratum cannot be identified until some survey data have already been collected, possibly from a previous iteration of a repeated cross-sectional survey or another similar survey. Further, this stratum may not be amenable to recruitment by a common design. This would create the basis for further division of the strata. These methods are similar to using partial *R*-Indicators in that they target under-represented regions of the covariate space. They differ in that they may also target regions in which there is wide variation in the survey outcome variable (a conditional mean based on the observed covariates) relative to other regions of the covariate space.

3.3.3 Simulation

Once strata have been identified, existing data can be used to test whether the identified strata are likely to meet the criteria specified earlier in the chapter. This can be done using a previous iteration of the survey or a previous wave's data. One approach is to eliminate each interviewed case, one by one, and then reweight the data to see how influential each case is on the estimates. However, the influence of a single case is usually not significant enough to differentiate it much from the rest of the sample. Instead, some strata must be identified which can then be tested for an influence on estimates. Here, the methods described earlier may be useful. If a large number of potential stratifying variables are suggested, then simulation can be useful for identifying a best subset.

Sanders et al. (2015) tried this simulation method for a survey of farming operations. They identified the strata by focusing on groups for whom there were plausible methods for increasing their response rates. For instance, response propensity models were used to identify cases that were difficult to contact but were not very likely to refuse. Such a group might have its response rate increased by spending more effort on contacting those cases. This method corresponds to the second criterion of identifying groups that may respond to a proposed strategy at a rate different (usually higher) than other strata.

Next, they simulated the impact of increasing the response rates for this group on the final, nonresponse-adjusted estimates. They found that relatively small changes in response rates for these groups did not generally lead to reductions in biases for the adjusted estimates. But drastic reductions in response rates overall did lead to biases. In this way, simulation revealed that the selected stratification did not meet the other criteria.

3.3.4 Other Methods for Creating Strata

The creation of strata is a complex task. More research is needed in this area. Methods from diverse fields such as control theory and operations research may be useful. A trial-and-error approach, informed by expert opinion, may also be able to identify useful stratifiers.

Methods from machine learning might also be used to define strata. Clustering algorithms have been developed that will identify groups that share similar attributes (see Hastie, Tibshirani, and Friedman [2009] for a discussion of methods). Two of the most widely used methods—K-means clustering and hierarchical clustering—may be useful for helping to define strata. These methods create subgroups that are homogenous with respect to some outcome. They may be particularly helpful when a large number of covariates or "features" are available to help define the subgroups.

3.3.5 Examples

We consider two examples: the NSFG and the Dutch LFS. The studies use a variety of auxiliary data. See Section 3.5 for a general discussion of data sources.

3.3.5.1 The National Survey of Family Growth

The NSFG experimented with using regression diagnostics to define strata. The auxiliary data used in this study include data from the sampling frame, paradata, and data from the screening interview (see Section 3.5 for a general discussion of data sources). Paradata include variables generated from records of each call attempt and interviewer observations made about neighborhoods, households, sampled persons (including estimated characteristics of these persons along with any relevant questions or comments they may have made). Specifically, interviewer observations included describing the housing unit structure (multi-unit structure vs. single family home) and whether the household is in a locked building or gated community. Interviewer observations are also made of neighborhoods, households, and—after screening—sampled persons. Call record data and contact observations were continually collected through all stages of interviewing.

The sampling frame data include data about the neighborhood (Census Block, Block Group, or Tract) from the 2010 Decennial Census. These data are very general and have been shown to be not very useful for predicting survey variables. Examples include the proportion in the ZIP Code Tabulation Area (ZCTA) that has never been married.

Since the screening interview collects the age, sex, race, and ethnicity of each person in the sampled household, these data are available as auxiliary data for households that completed the screening interview.

There are also data available from a commercial vendor. These data come from a variety of sources and are matched to the US Postal Services Delivery Sequence File. Match rates vary across the many variables available.

These match rates are extensively explored elsewhere (see Pasek et al., 2014; Sinibaldi et al., 2014; West et al., 2015). In general, about 70% of households have information such as the name of an adult household member and some information about ages of adults 18+ in the household.

There are several key statistics measured by the NSFG. For this study, we chose to use a binary variable, "never been married." This variable is collected for all respondents—men, women, and teenagers.

It is possible to use this approach to target influential cases across a range of key outcome variables. The sample design methods that have been developed for multipurpose surveys can be extended to the use of survey outcome variables in planning adaptive designs. The approach was experimentally implemented with a single outcome. The experiment was designed to be implemented in Week 6 of the 12-week data collection. At the end of Week 5, two regression models were estimated. The first model uses data from the respondents. Their reported survey outcome variable (never been married) is predicted using the available sampling frame, paradata, and screening data. The screening data are only available for households that have completed a screening interview. There was one version of the model estimated for cases that had not been screened, and another version for cases that had been screened. The latter models included data from the screening interview.

The predicted values and influence diagnostics for the responders were stored. Then, the influence diagnostics were split into two categories— "high" and "low." A second model was then estimated on the respondents predicting the probability of having high influence diagnostics. The estimated coefficients were then applied to the covariates of the current nonresponders to generate a score—the predicted probability of having high influence.

The remaining active cases were divided into two groups—experimental and control treatments. The experimental group had the 30% of cases with the highest predicted probability of being influential flagged. The flag was displayed in the sample management system. Interviewers were asked to make these flagged cases a high priority for interviewing. In other research, this prioritization scheme has been found to be effective in increasing the effort on the flagged cases. This extra effort frequently leads to higher response rates for the flagged cases. The control group had a random 30% of cases flagged.

Table 3.1 shows the characteristics of the persons interviewed under the experimental treatment (using regression diagnostics to identify a stratum of cases to be prioritized for additional call attempts) and under the control treatment (with random prioritization of cases). The first row of the table shows the estimate of the survey outcome variable for each group. From the results in Table 3.1, it seems that the use of regression diagnostics to select cases to prioritize leads to recruiting respondents who are not teens and who are currently living alone. This makes sense as it is relatively rare for teens to be married in the United States and single persons are not currently married, but may have been married. It is more difficult to predict the "never married" status for this group than for persons who are currently living with their spouse.

TABLE 3.1

Characteristics of Cases Recruited under Alternative Prioritization Schemes

Variable	Experimental	Control
Outcome: Pct Never Married	55.6%	53.8%
Commercial Data: Person 1 25–30	7.9%	8.8%
Commercial Data: Person 1 46+	39.7%	20.0%
Commercial Data: Marital = ?	6.4%	11.3%
Commercial Data: 1–2 Adults	11.1%	6.3%
Commercial Data: No one 18–24	68.3%	61.3%
Commercial Data: No one 35–64	34.9%	32.5%
Commercial Data: No one 65+	71.4%	62.5%
Not Self-Representing PSU	73.0%	73.8%
Single Family Home	52.4%	55.0%
Apartment	23.8%	23.8%
Sex Active Interviewer Observation	95.2%	91.3%
Screener: Hispanic	27.0%	23.8%
Screener: Race = Not White	63.5%	51.3%
Screener: Male	52.4%	47.5%
Screener: Single Adult HH	20.6%	11.3%
Screener: Teen	4.8%	25.0%
Segment Observation: No Access Impediment	71.4%	60.0%
Segment Observation: Poor/ Unimproved Roads	6.4%	13.8%
Screener: Age (mean)	28.6	26.5
ACS Block Group Percentage Never Married (mean)[a]	58.4	56.7
ACS Block Group Percentage of Population Age 15–44 (mean)[a]	0.3	0.4

[a] ACS=American Community Survey; Census Block Groups are areas that include about 600–3,000 persons.

A limitation of this approach is that it may not be easy to label the strata of interest. As a result, the design features may be effectively limited to prioritization such that targeted strata receive additional effort. A strength of this approach is that it includes survey outcome variables in the analysis.

3.3.5.2 Labor Force Survey

The LFS uses a sequential mixed-mode survey design. The sample receives a mailed invitation to respond by Web. Web nonrespondents receive a telephone follow-up whenever a phone number is available and otherwise a face-to-face follow-up.

A static ASD is considered in which the population is stratified beforehand based on linked administrative data. The five most relevant auxiliary variables to the LFS survey outcome variables are selected to form strata: a binary indicator for subscription to an unemployment office (registered as seeking a job), age, household size, ethnicity, and a binary indicator for registered employment (registered as employee). Crossing the five variables led to 48 population strata (yes or no registered unemployed in household times three age classes times two household size classes times two ethnicity classes times yes or no registered employment in household). The strata were collapsed to nine disjoint strata based on three criteria: differences in predicted Web response propensity, a differing propensity of having a listed phone number, and a large difference between estimates based on different multi-mode data collection strategies. All three criteria point to mode having an impact on estimates and mode-specific bias. Older households have very low Web response propensities, but high telephone registration propensities. Younger and non-native households have both low Web response propensities and low telephone registration propensities. Registered employed have higher Web response propensities, while registered unemployed have lower Web response propensities. The nine strata are

1. *Registered unemployed*: Households with at least one person registered to an unemployment office (7.5% of the population).
2. 65+ *households without employment*: Households with a maximum of three persons of 15 years and older without a registration to an unemployment office, without employment, and with at least one person of 65 years or older (19.8% of population).
3. *Young household members without employment*: Households with a maximum of three persons of 15 years and older without a registration to an unemployment office, without employment, with all persons younger than 65 years, and with at least one person between 15 and 26 years of age (2.4% of population).
4. *Non-western without employment*: Households with a maximum of three persons of 15 years and older without a registration to an unemployment office, without employment, with all persons younger than 65 years and older than 26 years of age, and at least one person of non-western ethnicity (1.5% of population).
5. *Western without employment*: Households with a maximum of three persons of 15 years and older without a registration to an unemployment office, without employment, with all persons younger than 65 years and older than 26 years of age, and all persons of western ethnicity (11.0% of population).
6. *Young household member and employed*: Households with a maximum of three persons of 15 years and older without a registration to an unemployment office, with at least one employed, with all persons

younger than 65 years, and with at least one person between 15 and 26 years of age (15.6% of population).

7. *Non-western and employed*: Households with a maximum of three persons of 15 years and older without a registration to an unemployment office, with at least one employed, with all persons older than 26 years of age, and at least one person of non-western ethnicity (3.9% of population).
8. *Western and employed*: Households with a maximum of three persons of 15 years and older without a registration to an unemployment office, with at least one employed, with all persons older than 26 years of age, and all persons of western ethnicity (33.5% of population).
9. *Large households*: Households with more than three persons of 15 years and older without a registration to an unemployment office (4.9% of population).

In Chapters 4 and 11, we will discuss the choice of strategies and optimization of these for the nine LFS strata.

3.3.6 Summary and Recommendations

Creating strata is a problem with several dimensions. It is important to keep this in mind while developing strata. Focusing on a single dimension may lead to suboptimal stratification. For example, there may be a temptation to select a response propensity model with strong predictors of response. Typically, paradata provide the strongest predictors of response. Yet, as the HSLS example shows, these predictors may not be useful when the goal of the model is to reduce nonresponse bias. Variables derived from level-of-effort paradata may have a strongly stochastic element that reduces their utility in relationship to this goal. For instance, the number of calls required to complete an interview is a complex (and varying) function of the other key elements of the survey design, the lifestyles of respondents, and multiple choices made by interviewers or via an automated sample management system.

The relationship to survey outcome variables, and ultimately to the possibility of bias after adjustment needs to be kept in mind. This might be done, for instance, by relying on a subset of available auxiliary variables that have correlations with key survey outcome variables. It will be useful to obtain input from subject area experts about useful predictors of key outcomes from the survey. If some data are available, testing these relationships is a good idea. If very few useful variables are available, identifying new sources of data is an additional necessary step. For example, it may be possible to construct interviewer observations tailored to the content of the survey.

Overall, identifying strata is a complex task. Given this complexity, it will be useful to try several different methods for creating strata. If multiple methods produce similar results, this is an indication that the results of an adaptive design are likely to be robust across several similar stratification schemes. If not, then it may be wise to consider why different conclusions are reached. It may also be useful to experiment with different stratification schemes to see if they produce discernibly different results. In the context of repeated cross-sectional designs, refining and testing strata over time may lead to improvements in outcomes. Given that this step is critical for the success of ASD, resources used in the evaluation of various stratification schemes will not be wasted.

3.4 Available Data

The previous section emphasized the importance of selecting the right variables for stratification for an adaptive design. This decision is limited to the variables that are available to any particular study. In this section, we review data that may be available across a wide variety of studies.

The available data are usually a function of the sampling frames. Therefore, we describe several commonly used sampling frames and give a brief description of the data that are generally available for each. These sampling frames include address frames (such as, but not limited to area probability frames), random digit dialing sample frames, and list frames such as population registers. These sampling frames can be supplemented with data from various sources. After discussing data associated with each of these sampling frames, we will discuss data that may be generated as part of the survey data collection process. These data are known as paradata.

3.4.1 Sampling Frames

Random digit dialing frames: Random digit dialing uses information about the telephone system to randomly generate telephone numbers. These numbers are generated using telephone banks that have been assigned for residential customers. In North America, these banks are generally assigned to landline or cellular services. It is possible, however, for users to move a number assigned to landline use to a cellular telephone and vice versa. Given that these numbers are randomly generated, there is very little information on the sampling frame. The data associated with these frames are generally added by linking data from other sources.

If the telephone number can be linked to telephone listings, then this linkage is a useful variable as persons with listed numbers tend to be more cooperative. Linking to the listing may also make it possible to add data from other sources (see below). If the telephone numbers can be assigned a local geography, either through linkage to directory listings or via knowledge of the geographic assignment of telephone numbers, then it is possible to add contextual data about the local area. These data would include summary information about the local population from a census or other large survey. Typical variables might include the total population; population per square kilometer; proportion of the population in various subgroups defined by age, race, and sex; and median home value. These contextual data are often weakly associated with response and many survey outcome variables. One useful source of data is the Planning Database prepared by the US Census Bureau. This database contains a number of predictors drawn from the Decennial Census and the ACS, as well as the response rates to the mailed portion of the Decennial Census. The Census Bureau uses these data to create hard-to-count scores for each tract in the United States. In the United States, for mobile telephone numbers, there is no systematic way to identify the geography of a telephone number. Mobile telephone numbers may be associated with a billing center, but research has found that even being able to determine in which state the person with the number is likely to be living may be difficult.

Address frames (including area probability samples): In the United States, it is possible to sample addresses using a recently developed method called address-based sampling (ABS). This method relies upon commercial vendors who construct address lists of sufficient completeness that the US Postal Service will complete the lists using the delivery sequence file (DSF) that they maintain. These address lists, when supplemented with rural route emergency address lists, provide a very high rate of coverage for the household population in the United States.

As with RDD samples, the ABS frame does not contain much information about each unit. There is an indicator for whether the unit is seasonal or if it is unoccupied, that is drawn from the DSF. These address lists can be geocoded such that the Census geography (i.e., tract and block numbers) are added to the data file. The Census geography allows the user to add contextual data from the Decennial Census, the American Community Survey, and the Planning Database. The same kind of variables that can be attached to a telephone number with an estimated geography can also be attached to geocoded addresses. The utility of these data are often limited in that they are weakly correlated with response, and often weakly correlated with survey outcome measures.

Register frame: In some countries, National Statistical Institutes may have access to population registers. These registers are assembled for various purposes, including taxation, voting, and social benefits. When available, these registers provide a high-quality sampling frame. These may include variables such as age, race, household size, home value, and employment status. These variables may be useful predictors for response. They can also be useful for predicting some survey outcome variables.

3.4.2 Commercial Data

In the United States, some commercial firms have made available for purchase data about households. The data are assembled from a variety of sources, but primarily from companies that assemble consumer credit information. These data include information on household size; demographic information about household members including age, sex, race, and ethnicity; financial information such as household tenure (own vs. rent) and estimated household income; and voting records. These data are linked via the address, but not every household has this information available. A recent study found that more than 80% of addresses in a sample of the U.S. population can be matched (West et al., 2015).

Commercially-available data are also somewhat inaccurate. These inaccuracies may be due to mobility or incorrect linkages. These imperfections reduce the utility of these data. Still, they can be useful for identifying households containing eligible persons, and predicting the likelihood of response. In particular, an indicator for whether these data are available for a housing unit is often a good predictor of likelihood of response (see West et al., 2015).

3.4.3 Paradata

Paradata were defined by Couper (1998) to mean data generated by the process of collecting data. Examples include call records and keystroke data. Over time, this definition has been expanded to include other forms of data that are not necessarily a by-product of the process. For example, interviewer observations may be collected solely for monitoring and adjustment purposes. These have also been included in the definition of paradata. Paradata, in this broader sense, include call records, keystroke or audit trail data, Web paradata, interviewer observations, and time and expense record data (Kreuter, 2013). We will give brief descriptions of two types of paradata: call records and interviewer observations.

Call records: Call records are generated by interviewer-mediated surveys. These records include information about each attempt: the time, date, interviewer, and result of the attempt. Interviewers may also leave detailed notes that are helpful for follow-up calls, but difficult to analyze since they are text. The results are usually recorded in some coding scheme that provides information about whether there was any contact, any refusal, appointments set, interviews taken, and any situations that prevent an interview from occurring—for example, the sampled person is unavailable.

These data have been used for monitoring purposes. For example, interviewer-level contact rates may be monitored to determine if an interviewer is having difficult scheduling call attempts at appropriate times. They may also be used for model predictions about when a call attempt is likely to lead to contact or an interview. Variables constructed from call record data may also be used for adjustment purposes. Although, for this purpose, there is mixed evidence about the utility of these data. In general, if these level-of-effort measures are not related to the survey outcome measures, then they may not be useful for adjustment purposes. Although strongly predictive of response, they say very little about the differences between those who respond and those who do not, with respect to the survey outcome measures.

Interviewer observations: Interviewer observations are made by interviewers. These can be observations of sampled neighborhoods, housing units or persons. These observations should be made in a similar way for all sampled units. For example, observations of sampled persons should all be taken when the interviewer has the same information about each person. That is, the observations should all be taken before an interview is completed. Otherwise, the observations about the respondents (which could be informed by the content of the interview) could systematically differ from those made about the nonrespondents (which could not be informed by the content of the interview). This difference could lead to biased inferences about the sample as a whole, since nonresponders might be systematically different with respect to the observations.

Examples of observations include whether there are any physical impediments that block access to the neighborhood, safety concerns, the condition of the housing unit or whether members of the household made any comments or asked any questions. These data are frequently useful for predicting response, but may be only weakly associated with survey outcome measures.

One advantage of interviewer observations is that they may be tailored to the survey content. For example, the NSFG asks interviewers to estimate (before completing the interview) whether the sampled person is in an active sexual relationship with a person of the opposite sex. This observation is correlated with many of the key statistics collected by the NSFG. In another example, from a study of the impact of a hurricane on the mental health of a local population, interviewers were asked to observe the level of damage to sampled housing units.

3.5 Summary

The goals of stratification for adaptive design are

1. Create strata that are influential on estimates.
2. Create strata that respond to different recruitment strategies in similar ways.
3. Create strata such that at least some strata may be given lower cost strategies.

Identifying the optimal stratification is a difficult balancing act. In many cases, the variables that are available for defining groups may be more useful for discriminating among cases that are easier or more difficult to interview. Certainly, level-of-effort paradata are very useful for this purpose. However, it may be more difficult to identify groups that will reduce the bias of adjusted estimates if their response rates can be improved relative to those of other groups. A useful starting point is to *identify predictors of the survey outcome variables.* If these do not exist on the sampling frame, finding a way to build these into the data—either as interviewer observations or from other sources—is an important step in the process. This requires more effort and will involve substantive area experts in the discussion. But this will be necessary in order to build useful stratification schemes.

No single set of tools for the creation of strata has been identified. Several example approaches were described in this chapter, including using Partial *R*-Indicators, particularly when based on estimated response propensities from models that include predictors of the survey outcome variables, and regression diagnostics. Other methods, borrowed from other fields, may also be useful—including clustering techniques from machine learning. Identifying strata may take a trial-and-error approach that improves the stratification over time. Although difficult, the empirical evidence points to this approach being useful for controlling nonresponse bias, even with suboptimal stratification schemes. Starting slow and building with experience is a useful strategy.

4

Interventions and Design Features

4.1 Overview

In an adaptive design setting, a key element is the design features that may be altered or tailored as part of the design. The literature on survey methodology has explored many possible aspects of survey design to see their impact on response rates. Some of the available design features have a strong effect on response rates, while other features have only a weak effect. For example, changing the mode of the survey request can have a strong impact on response rates while changing the stamp on a mailed survey request may have only a weak impact.

In addition to being concerned about design features that may affect response, we are also concerned with design features that differentially impact subgroups of the population. These offer the opportunity to target subgroups in order to reduce either errors or costs. Of particular interest are subgroups that differ with respect to the items being measured by the survey. This is, of course, a function of each particular survey and such groups may be difficult to identify ahead of time, particularly when data on the sampling frame or paradata are only weakly predictive of these items. For example, assuming that age is predictive of survey outcome variables, and if it is the case that younger persons are more likely to respond via the Web while older persons are less likely to do so, it would make sense to offer this mode only to young persons. This could reduce costs without harming response among either the older or younger groups.

Given the large number of available design features, in this chapter, we will focus on design features that have a relatively strong impact on response and that may have differential effects across subgroups in the population.

There are at least five aspects of each design feature that need to be discussed.

1. The design feature itself (e.g., incentive, mode change, etc.)
2. The dosage
3. The sequence

4. The empirical impact of the design feature on nonresponse bias
5. The costs

While it could be argued that the first aspect might subsume the others, we choose this delineation in order to highlight two oft overlooked features of each intervention. The dosage may be a critical component of the effectiveness of the design feature. Although not every feature will have different doses, many will, and this dosage can be an important feature. As an example, consider the design feature "additional call attempts." The effectiveness of this design feature depends greatly upon the dosage—especially the number of attempts. Each additional attempt adds to the final response, albeit with declining returns. In this case, the dosage is a critical component of the intervention. Often, particularly for mixed-mode surveys, the emphasis is on the mode change, while the dosage of each mode is seen as less important or even not reported.

The sequence is also an important, but overlooked aspect. The most studied impact of sequence is for incentives, with many experimental comparisons of pre- and postpaid incentives existing in the survey methodological literature. For other interventions, in particular, mode switching, the impact of sequence is less understood. In an adaptive design setting, the sequence can be an important feature. Optimizing adaptive designs is often the search for the optimal sequence of treatments across cases. The sequence can matter when there are interactions between the treatments across the phases of the adaptive design. For example, starting with a telephone mode followed by face-to-face will likely be more cost-effective than starting with face-to-face. However, if telephone attempts are used in the first phase, and they anger the sampled unit, they might reduce the effectiveness of face-to-face contact attempts. If the face-to-face attempts had been made first, then they might have elicited more response. In this case, the sequence of treatments does matter. In this case, it might be useful to target cases with the most effective treatment in order to avoid these effects. Recent evidence from surveys offering a sequence of mode is mixed on the impact of the sequence (see Wagner et al., 2014a and Tourangeau, 2017 for summaries). Some studies have found no effect. For example, McMorris et al. (2009) found that the two sequences of internet and face-to-face interviewing produced comparable response rates and estimates. In a recent study Dillman et al. (2009) compared several mixed mode designs and found that mail followed by telephone performed as well on response rate as telephone followed by mail. Other studies have found that the sequence does matter. For example, Beebe et al. (2007) compared response to a Web survey of physicians followed by mail to a mail survey followed by Web. They found that the Web/mail sequence had a lower response rate (62.9%) compared to the mail/Web sequence (70.5%). Smyth et al. (2010) found that a mail survey with Web follow-up produced higher response rates (71.1%) than a Web survey followed by a mail survey (55.1%). Of course, dosage may also play a role, but these studies are at least suggestive.

On the other hand, where these kinds of negative synergies do not exist, then switching design features may be conceived of as a search across alternatives, looking to identify the most effective treatment among all alternatives. In this case, the sequence does not matter, other than from a cost point of view, and may lead to improvements in response. Ideally, the design features in any sequence would be "complementary" in the sense defined by Groves and Heeringa (2006)—that is, different interventions are more successful with different strata, but these differences across the design features cancel out over the whole sequence with respect to the strata and produce balanced response rates across the strata.

Other aspects include the empirical impact of the design feature on nonresponse bias. This is a particularly important component since information about how the intervention works across the strata used for the adaptive design is essential for planning. Assembling evidence about nonresponse bias is a difficult task—particularly when focusing on strata. Much of the existing research is on overall estimates. In the simplest case, it is helpful to have estimates of response rates for each stratum under the set of designs that are being considered. For example, if two designs are being contemplated such as Web and telephone, knowing the predicted response rates for each of the strata under each of these two designs are important inputs to the adaptive design. In the best case scenario, there would be information about biases for each stratum under the different design options. This best case scenario is likely to be achieved only in very rare cases where gold standard data are available.

The costs associated with each design feature are also important. The survey budget limits the range of available options. For instance, some surveys may not be able to afford to include interviewers. In the case of adaptive designs, different treatments may be allocated across the strata based on the objective of maximizing a balance indicator (such as the R-indicator). For example, this might lead to allocating a more expensive design, such as face-to-face interviewing to strata where this will have the largest relative impact on response while allocating less expensive design features, such as a Web survey, to strata where the difference in response between Web and face-to-face modes is smaller. When design features are offered in a sequence, then costs may justify using subsampling at each phase in the survey in order to efficiently deploy costly design features. Survey costs will be discussed briefly in this chapter and more fully in Chapter 6.

One problem that can arise in practice is that there is not sufficient data available before the field period to assign cases to targeted treatments. For example, it may be that we cannot identify cases based on information on the frame for whom a face-to-face contact attempt is more effective than a telephone contact attempt. In this case, offering a sequence may be the only way to identify cases that need a face-to-face contact attempt. They will be unsuccessfully attempted via telephone and then will receive a face-to-face contact attempt.

In Chapter 7, we discuss methods for optimizing ASDs. In this chapter, we will look at a list of design features that could potentially be used in adaptive designs. These are, in some sense, the raw materials of these optimization problems. We will also specifically examine questions related to dosage and sequencing of these design features.

4.2 The Interventions

Many different interventions are possible. Not all of them can be discussed here. Rather, we focus on a subset that are important and have been shown to be useful for ASDs. Some of the key types of interventions include allocation of scare interviewer effort through case prioritization, varying the contact protocol, incentives, using multiple modes of data collection, and rules for stopping data collection for some or all sample members.

One frequently used approach is to allocate limited resources to cases that are deemed important on some criterion. This might be because of the predicted values for the survey questions, likelihood of response, or costs. A common way of allocating these scarce resources is to prioritize particular cases. How cases are "prioritized" can vary depending upon the mode of the survey.

In a face-to-face survey, prioritization can mean that the importance of the case is communicated to interviewers. The interviewers can be trained to either call these cases more frequently or to call them at times when contact is more likely to occur. This strategy was implemented by the NSFG (Wagner et al., 2012). In the NSFG, cases were prioritized based on imbalances in response rates for subgroups for variables that are fully observed. For instance, among identified eligible persons, subgroups based on age, gender, race, and ethnicity that had lower response rates were prioritized. While this prioritization frequently led to increased calls on cases that were prioritized (7 of 16 experiments), it only infrequently led to increases in response rates for the prioritized group (2 of 16 experiments). Obtaining interviewer compliance with these requests is an important prerequisite for implementing these types of designs in face-to-face surveys and may be a nontrivial task. Walejko and Wagner (2015) report on difficulties in implementing these types of interventions in a face-to-face environment. A more detailed examination of the NSFG example is given at the end of this chapter.

In a telephone setting, these prioritization schemes can be implemented through a sample management system, thereby negating the interviewer compliance issue. In a telephone setting, cases can be moved to the top of the list for calling, or called more frequently. Laflamme and Karaganis (2010) focus all the effort on a subset of cases at a specified point in the data collection. Wagner (2013) changes the order of cases to be called based on the previous history of calling each case and other covariates, thereby attempting to call each case at a time most likely to result in contact. Greenberg and Stokes (1990) used methods from operations research to identify a set of calling rules aimed at minimizing the number of calls needed to make contact. Bollapragada and Nair (2010), in the context of credit card collections calling, propose a strategy that searches a grid of call windows. The method uses overall average contact rates from each window as a baseline estimate. These estimates are updated with the results of calls made to the case. While strategies of this type may lead to increases in efficiency, there is little evidence that this translates into changes in estimates. Peytchev and colleagues (2010) compare changing other design features to increasing the number of calls and find that other changes in the design are more likely to lead to changes in estimates.

Another class of design features related to prioritization are those aimed at improving contact rates. These may include the timing of the calls (Wagner, 2013), but other features are available. For instance, in panel surveys, placing the first attempt at the time that the first interview was conducted has been found to improve contact rates (Lipps, 2012; Kreuter et al., 2014). Others have suggested changing the frequency of attempts (Weeks et al., 1987; Brick et al., 1996). However, this can be difficult as it is often the case that less frequent attempts are more likely to produce contact. Since surveys often operate under deadlines, this strategy may not always be feasible.

Mail surveys have their own contact strategies that involve the type, number, and frequency of mailings that are used. Dillman, Smyth, and Christian (2014) provide a summary of the "total design method" for implementing mail surveys. The method involves the dosage of the request to complete a mailed survey and follows a specified sequence of attempts: prenotification, a first mailing of the survey, postcard reminders, and second mailings of the survey. Recent research has examined the impact of offering different sequences of incentives across the various stages of the survey request (Dykema et al., 2015). Others have examined the use of a two-phase mailed survey for screening the general population in order to identify a subgroup of interest (Brick et al., 2011b, 2012; Montaquila et al., 2013). As difficulties with telephone surveys increase, interest in mailed methods may continue to develop.

Web surveys may combine several types of contact strategies based upon the information available on the sampling frame. This may include mailed notification for a sample of addresses (Smyth et al., 2010) and may also include email contacts (Klofstad et al., 2008). When both mail and email options are available, finding the appropriate sequence and number of each type of contact is an important question (Munoz-Leiva et al., 2010; Millar and

Dillman, 2011; Kaplowitz et al., 2012). Other research has looked at whether varying the content of the requests has an impact on response (Kaplowitz et al., 2012; Sauermann and Roach, 2013; Mavletova et al., 2014). In the adaptive design framework, there are two sets of decisions related to these contact strategies. First, can targeted contact strategies aimed at particular strata be identified? For example, faculty members at a university are more likely to respond to a survey request that includes a mailed letter rather than just an email request (Dykema et al., 2015). Second, what are the specific sequences and mixes of contact modes that work best? Millar and Dillman (2011) found that using multiple modes of contact was a useful strategy. However, little is known about specific mixes or sequences and their impact on subgroups in the population.

Incentives are an important design feature and one that has been found to consistently improve response rates (see Singer and Ye, 2013 for a review). There is less research on when incentives can effectively combat nonresponse bias. Incentives have been used in adaptive designs in several ways. Incentives have been varied across phases of a design (Wagner et al., 2012, 2017; Chapman, 2014). In this way, the early phases may identify persons for whom the topic of the survey is less salient. The leverage of the survey can be increased for these persons by increasing the incentive. However, there might be some persons for whom offering the lower incentive first creates intransigence. This may lead them to become nonrespondents when they would have responded had the higher incentive been offered first. If this is the case, the sequence of offering incentives is important.

The use of targeted incentives has received some attention. Targeted incentives might be accomplished through a phased design that does not initially use incentives, but then deploys them for late responders. Or, a targeted use of incentives might be developed using existing data. Trussell and Lavrakas (2004) discuss the use of differential incentives across subgroups. These differential incentives are based on a set of large-scale experiments testing different amounts. Earp and McCarthy (2011) discuss the use of response propensity models for identifying subgroups of establishments (farming operations) that would respond without an incentive, thereby suggested which groups could be targeted with incentives.

Perhaps one of the most powerful design features is the mode, especially the mode of interviewing which tends to have large effects on costs and response. The assignment of mode is a powerful feature for adaptive designs. Calinescu et al. (2013) explore the assignment of mode based on characteristics of the case. They use an operations research approach that minimizes the R-indicator for a fixed budget. A similar example is given at the end of this chapter. Chesnut (2013) identifies an important subgroup of the population that is extremely unlikely to respond to the American Community Survey (ACS) via the Web. The ACS sequentially deploys Web, mail, telephone, and face-to-face modes. Chesnut's results suggest that skipping the Web mode for a particular subgroup in the sample may reduce costs and possibly increase response rates by reducing the number of contacts. Borkan (2010) compared a Web-mail mixed mode design to a mail-Web design and found that the mail-Web design obtained a higher response rate and that the respondents were generally younger. Further work is needed to identify which combinations of modes work for specific subgroups in the population.

An important consideration in the use of multiple modes is whether different modes have an impact on the responses to the survey. There is a considerable literature showing that mode can impact the response process. One of the main ways in which mode can impact response is through the presence of an interviewer. This can lead, for some types of questions, to social desirability bias (Adams et al., 2005; Karp and Brockington, 2005; Tourangeau and Yan, 2007; Kreuter et al., 2008; Heerwegh, 2009). Balancing these potential impacts on measurement against both potential nonresponse errors and costs is an important area of research going forward. These issues are discussed more fully in Chapter 11.

The interviewer is an important feature of the survey design. Interviewers generally vary in their ability to recruit and interview respondents (Campanelli et al., 1997; O'Muircheartaigh and Campanelli, 1999; Purdon et al., 1999; Pickery and Loosveldt, 2002; Durrant and Steele, 2009). Switching interviewers has long been a strategy used in refusal conversion. Recent adaptive designs have explored assignment of sample to interviewers as a means for improving response. Luiten and Schouten (2013) preidentify cases that have a lower than average probability of response. These cases are assigned to the best-performing interviewers. The cases with higher than average probability of response are assigned to all interviewers. Peytchev et al. (2010) used an interviewer incentive to target cases that were deemed to be difficult using predictions from response propensity models estimated from previous iterations of the data collection. Other interviewer manipulations include alternate interviewer training protocols (Groves and McGonagle, 2001) and matching of respondent and interviewer characteristics (Dotinga, et al., 2005; Johnson, et al., 2000; Webster 1996).

An oft-ignored feature is the length of the questionnaire. Longer questionnaires have been associated with lower response rates (Dillman et al., 1993; Edwards et al., 2002; Deutskens et al., 2004; Rookey et al., 2012; Beckett et al., 2016). For many surveys, this option may be difficult or even impossible to implement. A few studies have tried to use a shortened questionnaire as a nonresponse follow-up strategy (Lynn, 2003; Roberts et al., 2014). Peytchev et al. (2009) used a reduced questionnaire as a follow-up strategy while changing several other components of the design, including a prepaid incentive and a higher conditional incentive. They found that this change in protocol led to changes in five of 12 estimates. Some studies have combined a change in the mode with a reduction in the survey length (Montaquila et al., 2013). Matrix sampling, where some sections of the survey are given to random subsets of respondents, are another option for reducing the length of surveys (Raghunathan and Grizzle, 1995).

Panel studies include some special design features not available to other types of surveys. Fumagalli et al. (2013) examine whether using tailored between-wave mailings helps improve response. They also explore between-wave methods for updating contact information. Couper and Ofstedal (2009) suggest the use of location propensity. They use logistic regression models to predict the probability that each case will be located. Based on these estimates, a subset of cases with low estimated probabilities of being contacted can receive additional locating effort. McGonagle and colleagues (2011, 2013) have also explored specialized locating methods, which could potentially be tailored to the difficult cases. A problem is that methods for updating contact information tend to be more effective with more stable households, that is, those for whom the updates are less useful.

Finally, an important decision for any survey is when to stop collecting data. This decision has frequently been based upon sampling error—that is, data collection runs until a specified number of interviews or overall response rate (given a fixed sample size, this amounts to roughly the same decision rule) has been achieved. Such a decision rule ignores potential biases that may occur. It may even encourage data collectors to interview cases that are relatively easy and, therefore, similar to cases that have already been interviewed. An adaptive design can look at the decision to stop working a case as a function of the potential for bias weighed against the costs of continued work. There is a small body of literature regarding stopping rule for surveys. Rao, Glickman, and Glynn (2008) propose several stopping rules. Some of these rules look at the expected biases of stopping using imputation. Wagner and Raghunathan (2010) propose a similar rule. Lundquist and Särndal (2013) develop an approach that seeks to balance response subgroup response rates. This is done by stopping effort on cases with relatively high response rates. In their example, they show that this leads to reductions in the nonresponse bias of adjusted estimates. Särndal and Lundquist (2014b) provide a conceptual reasoning for why this might be expected to occur. These stopping rules do not lead to targeted stopping of particular cases, which may be a useful avenue for adaptive designs.

4.3 The Dosage

Dosage is a sometimes forgotten aspect of particular survey design features. For example, incentives are essentially continuous in terms of dosage. Many incentive studies are explicitly looking for an optimal dosage, or amount, to offer (Singer and Ye, 2013). Trussell and Lavrakas (2004), for example, try increments of US$1 from US$0 to US$10. Mercer and colleagues (2015) provide a review of incentive studies that looks at the relationship of the amount of the incentive and response rates across multiple modes. They find that the relationship is not linear, with diminishing returns for higher incentives.

Dosage is relevant for other aspects of the survey design as well. In mixed-mode surveys, the number of attempts to recruit sampled units to each mode is a relevant design feature. Many studies have compared mixed-mode strategies. However, without including description of the dosage for each mode, it is difficult to assess whether the sequence or specific dosages of each mode is more responsible for the final results. Wagner et al. (2014a) explore some of the difficulties with controlling the dosage of a face-to-face mode.

The length of the questionnaire is another nearly continuous design feature that can be offered in different doses. We previously discussed the impact that length of interview can have on response, and how reducing the length of the questionnaire might be a useful nonresponse follow-up procedure. Beckett and colleagues (forthcoming) use the results from a large number of patient satisfaction surveys, which have similar content and designs, to estimate the impact of the length of the survey on the response rate. They found that adding 12 items was associated with response rates that were 2.5% points lower.

For each of these examples, identifying the appropriate dosage can be difficult. Not all surveys may have the wherewithal to try 11 different incentive conditions. The problem is further amplified for adaptive designs that may want to use a targeted procedure for subgroups in the population. Clinical trials have confronted these issues in attempting to identify the optimal dosage for medicines. Berry (2001) discusses Bayesian approaches to dose-finding studies. Thall and Lee (2003) describe two methods for identifying the optimal dose. Thall and Russell (1998) examine a method for finding the optimal dosage of a treatment when both efficacy and toxicity need to be considered. These methods may help adaptive designs efficiently allocate resources across a range of options. For example, testing incentives may require testing a wide range of options. Previous research has found that there are nonlinearities in the relationship between the amount of the incentive and either the response rate or the cost per complete (Kristal et al., 1993; Andresen et al., 2008; Romanov and Nir, 2010; Mercer et al., 2015). Identifying the optimal amount may include testing a range of options. Methods from clinical trials may provide an efficient way to implement this type of experiment, where the allocation is adaptive based on early results, and locates the optimal amount relatively quickly.

4.4 The Sequence

Another feature of survey design is the sequence in which treatments are offered. In the broadest sense, each sequence could be thought of as a unique treatment. Then it is possible to reduce the complexity by identifying situations where the sequence does not matter. For example, if the timing of an incentive offer matters such that offering it before making contact attempts is more efficient than offering it only after making contact attempts, then the sequence matters. If the sequence does not matter, than the incentive can be viewed in isolation from the contact strategy.

Are there instances where the sequence does matter? There are several common examples from the existing survey literature. For example, prenotification is an example of a contact attempt that has been shown to be useful in improving contact and interview rates. It has to happen before other attempts in order to be effective. Refusal conversions are another example. They only occur after a refusal has been taken. A less obvious sequencing problem is incentives. In general, most experiments have found that prepaid incentives are more effective at inducing response and even, in some cases, more cost efficient (Singer and Ye, 2013). In these cases, the sequence of the incentive and other steps in the process has an important consequence.

Another example of this type of interaction between treatments that indicates the sequence is an important feature of the design comes from research into mixed-mode surveys. Smyth et al. (2010) explore the impact of the sequence and "dosage" of modes on response rates. They consider sequences of two modes—mail followed by choice of Web or mail, and Web followed by choice of mail or Web. In each case, the first request to complete the survey included only the Web or mail option. The first request was followed by a reminder postcard for nonresponders. This is an additional "dose" of the first mode. The mode switch was a single mailing that included both the URL for the Web survey and a paper survey with return envelope. In the case where a mail survey request was followed with a request to do a Web or mail survey, the response rate was 71.1%. When the sequence was reversed, Web request followed by mail or Web, the overall response rate was only 55.1%. It appears that in this case, sequence and dosage do matter. Offering the Web option first seems to reduce the ability of the mailed survey as a second mode to improve the response rate. On the other hand, it is possible that the mail mode is more effective and the increased exposure to this mode leads to higher response rates. Further research is needed employing a full crossover design, with equivalent dosage of each mode, to determine if the sequence of modes by itself interferes with response rates for a second mode.

In another example drawn from a mixed-mode survey experiment, Wagner and colleagues (2014a) show that the sequence of modes used for a screening survey (i.e., a survey used to determine whether a household contains eligible persons) had an impact on response to a request to complete a detailed interview. The request for the detailed interview did not vary across the experimental arms. In this case, an earlier treatment appears to have had an impact on the effectiveness of later treatments.

Designing the appropriate sequence of treatments may be difficult. If each sequence is thought of as a treatment, then a large number of treatments are possible and sample size becomes an issue in estimating the impact of sequence. Collins and colleagues (2007) discuss strategies for identifying useful sequences of treatments and designing efficient experiments to test these. A step that all surveys can take would be to report sufficient detail on survey experiments to see the full context, including potential sequences of treatments. This would help others evaluate whether sequencing matters for particular kinds of treatments.

In other cases, it may be useful to adopt methods from other fields, such as machine learning, that aim at optimizing sequences of decisions. Bather (2000) provides a useful introduction to methods for optimizing sequential decisions. Sutton and Barto (1998) describe methods for learning how to maximize specified quantities. This may require taking actions aimed at reducing uncertainty (i.e., exploring options that are less understood) rather than exploiting options that are more fully understood.

Related to the question of sequence is the concept of multi-phase sampling for nonresponse (see Hansen and Hurwitz, 1946). In the case of two phases, this is sometimes known as two-phase sampling. The approach is to apply a design strategy to the sample, then take a random subsample of those who do not respond and apply a new design to that subsample. This can be efficient when the designs grow increasingly expensive across the phases. That is, the relatively inexpensive early phases are applied to a larger group, while the more expensive later phases are applied to smaller groups. In this way, it is akin to optimal allocation for stratified samples. There is, however, the constraint that the sample at each phase is smaller than the sample at the previous stage. An example of two-phase sampling from the NSFG is discussed in the next section.

Table 4.1 summarizes information on interventions discussed in this chapter. For each intervention, relevant citations from the survey methodological literature are given, along with some indication about any empirical associations (positive, negative, or none) with nonresponse error, measurement error, and costs.

4.5 Examples

4.5.1 The National Survey of Family Growth

The NSFG uses several adaptive design strategies. We will describe the following three interventions:

1. Case prioritization to reduce variation in subgroup response rates.
2. Phased design with design features designed to be complementary across the phases.
3. Management intervention with interviewers who perform outside of specified bounds in administration of the questionnaire.

TABLE 4.1

Interventions, Relevant Literature, Impact on Nonresponse, Measurement Error, and Cost

Intervention	Relevant Review Literature	Impact of Intervention: Nonresponse Error	Measurement Error	Cost
More Contact Attempts		Reduces nonresponse, tends to recruit persons similar to those already recruited (Peytchev et al., 2009)	May have an association if later recruited responders are poor reporters; mixed evidence on this (Olson, 2006; Hox, De Leeuw, Chang, 2012)	Each attempt has a marginal cost which may vary greatly depending upon the mode (Gfroerer et al., 2002)
Incentives	Review: Singer and Ye, (2013); Targeted incentives: Earp and McCarthy (2011); Trussell and Lavrakas (2004)	Much empirical research of impact on nonresponse; limited research on nonresponse bias (Groves, Presser, and Dipko, 2004)	Incentive may change the mood or disposition of the respondent, may also encourage careful reporting	Costs vary, but nonlinear impact on response rates (Mercer et al., 2015)
Prioritization		May lead to more balanced response (Wagner et al., 2012)	Unknown	Normally a reallocation of effort within the same fixed budget
Interviewer Assignment		Assigning the highest-rated interviewers to the most difficult cases in telephone samples increases sample balance (Luiten and Schouten, 2013)	Assigning interviewers similar to respondents may reduce social desirability bias (Webster, 1996; Johnson et al., 2000; Dotinga et al., 2005)	Costs similar to standard protocol

(Continued)

TABLE 4.1 (*Continued*)

Interventions, Relevant Literature, Impact on Nonresponse, Measurement Error, and Cost

Intervention	Relevant Review Literature	Impact of Intervention		
		Nonresponse Error	Measurement Error	Cost
Use of Multiple Modes	De Leeuw (2005)	Use of multiple modes may reduce nonresponse error; sequence can be important for response rates; subgroups respond differently to different modes (true for sequences?)	Measurement properties vary across modes; social desirability acts differently (Kreuter, Presser, and Tourangeau, 2008); presence of interviewer makes a big difference (Klausch et al., 2013)	Costs of modes vary widely; mixed-mode surveys have higher fixed costs
Mailed Survey Contact Strategies	Dillman, Smyth, and Christian (2014)	Design of contact strategy is critical to response	May have an association if later recruited responders are poor reporters; mixed evidence on this (Olson, 2006; Hox, De Leeuw, Chang, 2012)	Mailed contact attempts, with or without surveys, may vary in cost
Web Survey Contact Strategies	Couper (2008)	Mixed-mode contact strategy may be useful; mailed paper prenotification may be helpful (Millar and Dillman, 2011; Kaplowitz et al., 2012)	May have an association if later recruited responders are poor reporters; mixed evidence on this (Olson, 2006; Hox, De Leeuw, Chang, 2012)	Electronic and paper modes of contact very different in costs; multiple modes of contact have higher fixed costs

(*Continued*)

TABLE 4.1 ***(Continued)***

Interventions, Relevant Literature, Impact on Nonresponse, Measurement Error, and Cost

		Impact of Intervention		
Intervention	**Relevant Review Literature**	**Nonresponse Error**	**Measurement Error**	**Cost**
Stopping Rules: Survey-Level		Several rules have been designed to limit risk of nonresponse and rely upon imputation (Rao, Glickman, Glynn, 2008; Wagner and Raghunathan, 2010); others have developed empirical rules aimed at minimizing nonresponse bias (Lundquist and Särndal, 2013)	Unknown	Rules have focused on bias, but possible to add a rule based on variance; for imputation-based methods, costs are uncertain since stopping is a function of respondent data; empirical methods seek cost savings
Stopping Rules: Case-Level		Case-level stopping rules are discussed in Chapter 7; the rules seek to minimize the risk of nonresponse bias	Unknown	Cost savings are a goal

4.5.1.1 Case Prioritization

The first intervention is implemented when the response rate of particular subgroups lag behind the overall average response rate by a large amount. At that point, cases in the subgroup with the lagging response rate are prioritized for calling. This prioritization appears as a flag next the sampled unit's record in the computerized sample management system. Interviewers are instructed to pay special attention to these cases—calling them first, more frequently, and at better times within the constraint of continuing to organize their work efficiently. This prioritization led to a statistically significant higher response rate for the targeted subgroup in 2 out of 16 experiments (Wagner et al., 2012), but the pattern of higher response rates for the prioritized group was consistent across all the experiments.

The choice of which groups to monitor was an important one in this application. There are certainly groups that can be identified that have low response rates. For instance, there are those that refuse to complete the interview during an initial request. The proportion that will later assent to do the interview is quite small. However, this characteristic—early resistance—is not associated with many of the variables measured by the survey. Instead, subgroups were identified based on characteristics that were associated with many of the survey outcome variables—race, ethnicity, age, and interviewer observations about the selected person. A key interviewer observation was made after a person was selected to be the respondent but before that person was interviewed. Interviewers observed whether they believed the selected person was in a sexually active relationship with a person of the opposite sex. This observation, although measured with error, is associated with many of the key statistics produced by the NSFG (West, 2013). Therefore, these variables are useful predictors of key statistics in the survey and monitoring them for balanced response rates across their categories is useful.

4.5.1.2 Phased Design Features

A second set of interventions are bundled together as a package designed to produce a complementary recruitment protocol. There are two phases of production. The key design features of each phase are listed in Table 4.2.

TABLE 4.2

Key Design Features by Phase

Phase 1	Phase 2
US Mail Prenotification	FedEx Letter
No Screener Incentive	US$5 Prepaid Screener Incentive
US$40 Conditional Incentive for Main Interview	US$40 Prepaid Incentive plus US$40 Conditional Incentive for Main Interview
Large Workload	Small Workload

During the first phase, the relatively easy interviews are identified. Once these have been identified, at the end of phase one, a subsample of the remaining difficult cases is drawn and the phase two protocol is applied. The goal is to create a design such that the probability of an interview in the second phase under the second phase design is higher than the probability of an interview during the second phase under the first phase design. This can be accomplished by a variety of means, but typically involves a more expensive protocol.

An unanswered question is whether it is possible that the phase two protocol is less effective after the phase one protocol. A goal of adaptive designs is to identify subgroups for which this is the case. In order to do so, variables that predict a higher probability of response for the phase two protocol, when it is applied first, are needed. Also needed are experiments to test whether the probabilities of the two protocols producing response at phase one differ. Unfortunately, such experiments were not conducted.

If, on the other hand, the only variables that are able to identify the cases more likely to respond to the phase two protocol are the history of previous treatments, that is, they did not respond to the phase one protocol, then such a phased design may be necessary.

4.5.1.3 Interviewer-Level Management Intervention

The last intervention was conducted at the interviewer level. Paradata from the interview was monitored to identify potential problems during the interview. For example, these data include counts of the number of time there were backups, error escapes, suppression of error messages, don't know/refused responses, help keys were used, and the time each field takes. These data, when summarized to the level of the interview may indicate that a particular respondent had trouble with the questionnaire. When summarized to the level of the interviewer, then a pattern for each interviewer across multiple interviews may be observed. This may provide evidence that interviewers are having trouble or not following established interviewing protocols. Interventions to improve interviewer performance may be needed in this case.

For monitoring purposes, statistical summaries of these behaviors may be used. In the case of the NSFG, these errors are summarized into three factors: going too fast, too many error checks, and too many "don't know" and "refused" responses. Interviewers who are more than three standard deviations above or below the norm in each of these factors for three or more weeks in a row will receive an intervention. First, supervisors will review the details of each factor to see which specific behaviors are occurring at higher rates. Then the supervisor will meet with the interviewer to go over the issue. This may include a reminder of the established protocol for interviewing or the procedures for administering the survey questionnaire.

4.5.2 The Dutch LFS

The Dutch LFS is commissioned by Statistics Netherlands. Around 2008 the Board of Directors decided on the following general strategy for in-person and household surveys: All surveys start with Web data collection, possibly supplemented by mail; and interviewer modes (telephone and face-to-face) are employed only when accuracy is insufficient. When accuracy is insufficient, then either Web (plus mail) may be replaced by a mix of telephone and face-to-face, or a follow-up to nonrespondents may be added using a mix of telephone and face-to-face. In the following, we abbreviate the modes to W (Web), M (mail), T (telephone), and F (face-to-face).

As a consequence of this decision, the survey mode became the primary design feature to be tailored. Since the LFS questionnaire has many filter questions and a complex routing, the mail survey mode was not considered an option. In a series of pilot studies, five mode strategies were tested: W, T, F, W → T, and W → F. Here, "→" refers to a sequential strategy where nonrespondents to Web receive a follow-up.

Calinescu and Schouten (2015) optimized the choice of the mode strategy for a stratification of the population based on the Dutch Population Register and Dutch Tax Board Register. The stratification for the LFS was discussed in Chapter 3, and has nine strata based on age, size of the household, number of registered unemployed, and ethnicity. Since the number of calls in the interviewer modes influences both costs and response rates, they added this design feature to the optimization. A rough distinction was made between a standard effort (S) and an extended effort (E) in terms of the number of calls, leading to nine strategies: W, TS, TE, FS, FE, W → TS, W → TE, W → FS, and W → FE.

Tables 4.3 and 4.4 present the estimated response propensities per strategy and stratum and the estimated telephone propensities per stratum. The telephone propensity represents the probability that a phone number can be found in commercial databases. Table 4.3 shows that response propensities are very different over the nine strategies. Response propensities also vary over strata but to a lesser extent. Table 4.4 shows that telephone propensities vary greatly over the strata.

4.6 Conclusion

ASDs rely upon the ability to change the data collection protocols in ways that will change the behavior of sampled units. For different strata in the population, this might mean different protocols are optimal. Matching these protocols to strata is a difficult task. The focus has been on how these designs impact nonresponse across various strata. Other errors associated with these

TABLE 4.3

Estimated Response Propensities Per Strategy and Stratum

	Stratum								
	1	2	3	4	5	6	7	8	9
W	23.2% (0.2)	23.6% (0.2)	15.5% (0.2)	10.8% (0.2)	27.9% (0.2)	27.7% (0.2)	17.5% (0.2)	36.7% (0.2)	22.4% (0.2)
TS	12.2% (0.1)	31.4% (0.2)	8.5% (0.1)	4.7% (0.2)	19.7% (0.2)	13.3% (0.2)	7.2% (0.1)	18.1% (0.2)	21.2% (0.2)
TE	20.8% (0.2)	41.3% (0.2)	15.2% (0.2)	8.6% (0.2)	31.1% (0.2)	23.8% (0.2)	14.3% (0.2)	33.3% (0.2)	37.5% (0.2)
F	43.5% (1.0)	53.5% (1.0)	42.2% (1.2)	34.1% (1.0)	45.1% (0.8)	45.3% (1.0)	35.9% (0.8)	46.7% (1.0)	54.6% (1.0)
FE	52.4% (1.0)	58.3% (1.0)	51.0% (1.2)	41.2% (1.0)	51.2% (0.8)	54.9% (1.0)	46.0% (0.8)	56.8% (1.0)	61.4% (1.0)
W→TS	28.3% (0.2)	41.0% (0.2)	20.2% (0.2)	13.9% (0.2)	36.3% (0.2)	34.0% (0.2)	20.8% (0.2)	44.5% (0.2)	23.1% (0.2)
W→TE	32.8% (0.2)	48.4% (0.2)	23.8% (0.2)	17.5% (0.2)	42.1% (0.2)	41.1% (0.2)	25.8% (0.2)	52.1% (0.2)	24.4% (0.2)
W→FS	46.3% (0.7)	57.7% (1.3)	38.6% (0.7)	32.7% (0.7)	50.0% (0.8)	51.0% (0.8)	39.3% (0.6)	58.9% (0.9)	50.0% (0.4)
W→FE	49.8% (0.8)	58.3% (1.3)	43.4% (0.7)	36.6% (0.8)	52.6% (0.9)	54.7% (0.8)	44.3% (0.7)	62.0% (0.9)	54.2% (0.4)

TABLE 4.4

Estimated Telephone Propensities Per Stratum

Stratum								
1	2	3	4	5	6	7	8	9
38.1% (0.7)	76.4% (0.8)	30.2% (0.7)	22.4% (0.6)	60.0% (0.7)	38.9% (0.7)	32.0% (0.6)	53.4% (0.7)	62.4% (0.8)

designs should also be considered. For example, the mode of interviewing may have a large impact on the quality of the measurement.

In this chapter, we have given some examples of interventions and matching those interventions to strata. There are many other potential design features that could be considered for the adaptive design context. Future research should work to expand these options and identify subgroups to which they are well matched.

Further, when building these interventions, fitting them into a sequence is important. Of course, not every treatment will depend upon its location in a sequence of treatments. Given the focus of previous research on specific design features (e.g., incentives), focusing on potential interactions between treatments in a sequence may be a useful corrective. It should be considered when designing survey experiments.

An additional factor to consider is the dosage of each treatment. As with incentives, there may be interesting nonlinearities that should be considered. Understanding the impact of a range of possible dosages can be very important when design changes, as in the NSFG phased design, are being considered.

5

Models for Nonresponse in Adaptive Survey Design

5.1 Introduction

In design-based inference that underpins much of probability-based sample surveys, it may seem counter-intuitive that models can be used during data collection to improve the inferential properties of the survey data. Yet, implicit models have always been a staple in survey research and making them explicit can help to improve their use to meet study objectives.

Consider a survey on income and income inequality with two sampling strata, one of low-income areas and another of all remaining areas. A data collection manager would likely monitor response rates in each stratum. If the low-income stratum is yielding fewer interviews in proportion to the released sample, then more interviewing resources may be directed to the low-income areas. There is an implicit model in this example—sample yield by stratum.

Alternatively, the objective can be formalized. It can be to reduce the risk of nonresponse bias in estimates of income inequality. One could estimate the eligibility rate in each stratum that would be used to compute response rates. Even more directly, one could compute models to identify which cases contribute to nonresponse bias in each stratum and target them during data collection.

Models allow the use of a wider array of auxiliary information. For example, one could target sample members who have previously refused without a statistical model. A logistic regression, however, could incorporate other indicators associated with survey participation to improve the estimation of response propensities.

Particularly for models aimed at reducing nonresponse bias by increasing sample balance, the main risk is usually inefficient interventions and they seldom pose any risks of biasing survey estimates. For other objectives such as reducing cost, avoiding any increase in nonresponse bias becomes a priority in modeling. If one were to devise an intervention for the most difficult cases—identified by a logistic regression—it will likely have one of two outcomes. The model may have good fit and effectively identify cases with low response propensities, as intended. The model may also have poor fit and fail to sufficiently distinguish lower and higher propensities. In this latter case, the model would simply yield little benefit compared to using random sampling for the intervention.

Models are used in survey data collection, but making their use explicit can improve them and increase the benefit from their use. In this chapter we introduce the diverse uses of models for nonresponse in ASD, why they are needed, and their key components.

5.2 Goals of Statistical Models in ASD

A statistical model can be useful if its objective is clear. There are numerous goals that could be helped by models, such as reduction of cost, nonresponse bias, measurement error bias, variance, and nonresponse rates.

For example, a model could be used to identify cases in order to address nonresponse. This, however, is still not a clear objective. In order to reduce nonresponse at the lowest possible cost, the objective would be to identify nonresponding cases that are associated with higher response propensities and with lower cost to complete (e.g., lower travel cost). However, if the goal is to reduce potential nonresponse bias, the objective would be to identify nonrespondents who are different from the already interviewed respondents. If variance was another objective, then nonrespondent cases with large sampling weights may be targeted.

Most models can be characterized as having two major goals, error reduction (including all sources of error, both bias and variance, and proxy measures for error such as response rates) and cost reduction. This distinction in objectives is presented in the rows of Table 5.1. We will return to the examples later in this chapter.

Another distinction that was introduced in Chapter 2 is between RD and ASD, presented in the columns of Table 5.1. In RD, models are estimated at particular points in time with the intention to change the survey protocol for all or a group of sample cases. In an ASD the models identify individual cases that should receive a different treatment, often evaluated on an ongoing basis during data collection.

5.3 Models

5.3.1 Reasons for Using Models for Nonresponse

Estimation of nonresponse error almost always requires models—key survey variables are seldom measured on the full sample. There are several main reasons why the *explicit* use of models can be beneficial: they allow the

TABLE 5.1

Types of Models by Purpose and Implementation

		Implementation	
		By Phase/Group Level (Responsive Design)	**Continuous/Case Level (Adaptive Design)**
Main objective	Error reduction	Multiphase designs often with targeting of cases. Phases are implemented at a point in time. (Peytchev et al., 2010; Pratt et al., 2013)	Changing treatment on a case basis, during the course of the study (Calinescu, 2013).
	Cost reduction	Stopping rules for the entire study. Can be based on changes in estimates or imputed estimates (Wagner and Raghunathan, 2010). Targeting of more productive cases (Groves and Heeringa, 2006).	Attempt cases by propensity in a call window (Couper and Wagner, 2011). Mode switch (Chesnut, 2013). Stop cases based on propensity (Peytchev, 2013).

formalizing of data-driven decisions, they facilitate the use of a wider array of data, and they provide the means to balance multiple objectives.

Formalizing data-driven decisions. If a data collection manager is looking to identify underperforming interviewers, simple or implicit models such as directly comparing interviewer response rates can be uninformative or even misleading. A particular interviewer may have a low response rate, but may also be working in an area with low response rates. The manager will lack the criteria to determine which interviewers are underperforming. Suppose now that a statistical model is employed to control for sample member characteristics, including geographic area. The resulting model-based interviewer performance measure could be compared across interviewers and thresholds set to trigger action by the manager.

Leverage more data to better achieve goals. Data to inform decisions could come from multiple sources, such as

- Survey responses
- Survey data from prior survey iterations or waves
- Frame variables
- Auxiliary data from other surveys and censuses
- Administrative records and registries
- Real-time or near-real-time paradata
- Paradata from prior implementations
- Interviewer observations

These data are measured at different levels of geographic aggregation, with different periodicities, and different measurement properties. As a result, they may require the use of specialized statistical methods—such as multilevel models for nested paradata and survival models with time-varying covariates—but models are needed to account for the complex data structures created by the diverse types of auxiliary information to allow their use.

Balance multiple objectives. In the absence of models to estimate survey errors, resources may be allocated to the error source deemed to have the largest influence on survey estimates. Instead, models could be used to estimate errors from multiple sources, comparing their magnitudes, and allocating effort and resources so that survey estimates could be improved the most.

5.3.2 Key Components in Models for Nonresponse

Timing. Models can be estimated prior to data collection, during the current data collection, or a combination.

- *Prior to data collection.* In longitudinal surveys, data from the prior waves could be used for multiple objectives, such as identifying sample members who are less likely to participate. Prior wave nonrespondents tend to be less likely to participate in the following wave of data collection. Even those who participated in the prior wave but were less interested in the topic or said they were too busy, have been found to be less likely to participate in the next wave (Lepkowski and Couper, 2002). Peytchev et al. (2010) used prior wave data to identify which sample cases to target, and while the intervention was not effective, the model was effective in identifying cases with substantially lower response propensities.

  Prior data collection could be used even in cross-sectional surveys. In a telephone survey, Peytchev (2010) used a prior data collection to model the likelihood of an interview using census data aggregated to the sample member's census block group or ZIP code (the latter if an address was not available). The parameter estimates were saved, and applied to a new sample for the next iteration of the survey, to obtain estimated response propensities prior to data collection. Sample cases with low response propensities were randomly assigned to be called by an experienced interviewer until the first successful contact, under the expectation that the first contact may play a vital role in gaining cooperation from a sample member. Figure 5.1 shows that by assigning those cases to the more experienced interviewers, the interview rate among those cases was increased to about the same rate as the higher propensity cases.
- *During data collection.* Models could be estimated during data collection, often using survey data and paradata as they are collected. In this approach, often the dependent variable in a model is one

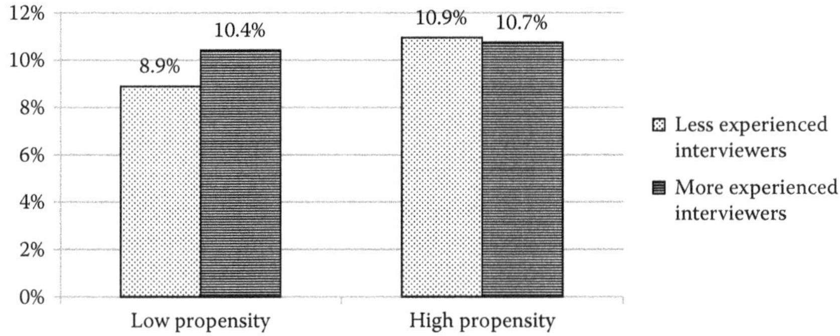

FIGURE 5.1
Interview rates by estimated response propensity and assignment to interviewer experience group. (Adapted from Peytchev, A. Responsive design in telephone survey data collection. International Workshop on Household Survey Nonresponse. Nürnberg, Germany, August 30, 2010.)

that is based on current data collection. For example, Wagner (2013) modeled the likelihood of an interview for four different call windows (time of day and day of week combinations) in a telephone survey and in a face-to-face survey, using aggregate census data, and updated those predictions during data collection. Through this approach, he managed to increase the contact rate in the telephone survey.

- *Hybrid combinations are also possible.* The models during data collection may be the most sensitive to the outcomes in a particular sample, but they may be unstable when based on few cases, particularly early in data collection. Furthermore, estimating a model very early in a multi-phase design may not yield the desired information—such a model may predict who would become a respondent early in data collection, yet the ultimate goal may be to predict who would remain a nonrespondent at the end of the entire data collection effort. Combining data from prior and current data collection efforts aims to balance the desirable properties of each approach.

  There are possibly two major types of hybrid approaches. One is when a prior survey is used to obtain the model parameter estimates, such as regression coefficients, which are then applied to the current data collection. This is the approach used by Peytchev (2013) to determine which cases in a telephone survey are unlikely to yield an interview by the end of data collection. A second, and promising, option is to estimate the needed model on data from the prior and from the current iterations of the survey, and gradually give more weight to the estimates based on the current data collection. There seems to be potential for applying a Bayesian modeling approach in this setting.

Variable selection. There are two sets of decisions related to the choice of variables: the dependent variable(s) and the covariates. Both need to match the primary objective. Increasing response rates, reducing the variability in response rates across subgroups, reducing the uncertainty resulting from missing information, and reducing estimated nonresponse bias, are all objectives related to nonresponse, yet the choice of variables to be included could be very different depending on the specific objective. The choice of a dependent variable needs to also match the intervention—if the intervention aims to reduce the likelihood of a refusal, then maybe the dependent variable needs to be refusal, rather than nonresponse in general. There may also be the need for multiple dependent variables, particularly when designing an intervention to reduce nonresponse bias in survey estimates.

In addition, the choice of covariates also depends on their source and measurement properties. Random measurement error in interviewer observations and random error in linking administrative data may merely reduce the effectiveness of any interventions based on such data, but such measurement error may have systematic components, inducing bias in the estimates. The choice of covariates also changes the meaning of results. For example, response propensities may be based on any model in which the dependent variable is an indicator of response to the survey. Yet, as Wagner and his colleagues have shown (Wagner et al., 2014b), including indicators of level of effort in the model increases the ability to differentiate between low and high propensities, but with little to no association with the survey estimates of interest.

Statistical methods. These can vary by purpose, data structure, and preference. By purpose, a model may need to produce the propensity of an interview at the case or call attempt level, as some of the earlier examples. However, when the purpose is to identify phase capacity, a point at which there may be no change in estimates and a change in protocol is ready to be started, nonresponse bias indicators are more appropriate, such as the fraction of missing information (Wagner, 2010) and propensity-based *R*-indicators (Schouten and Cobben, 2007).

Statistical methods can also vary by data structure. Models for survival analysis are particularly appropriate for call record data, multilevel models can be suitable for data with nested structure, and even both together (Steele and Durrant, 2011).

In addition, sometimes the choice of a statistical method may be influenced by preference—such as identifying sample cases using logistic regression, Mahalanobis distances, or *R*-indicators—as such choices seem to lack empirical evidence to support one statistical method over another.

5.4 Monitoring Nonresponse

There have been two proposed ways to classify measures of survey representativeness that could be more useful than the response rate in measuring and monitoring the potential for nonresponse bias. One classification is by whether the measure is at the survey level or at the estimate level (Groves et al., 2008). The other classification is by whether survey variables are used in computing the measure (Wagner, 2012). The two classifications are in near-complete agreement in practice.* For the ease of presentation, we will use the survey-level versus estimate-level distinction.

5.4.1 Survey-Level Measures

These measures share the desirable property of the response rate of being able to produce a single measure for the survey. They also share an underlying premise—for nonresponse bias to be present, response rates have to vary across subgroups that are homogenous with respect to the survey variables of interest. The latter part is challenging to ascertain, but summarizing the variability in response rates across subgroups goes beyond a simple response rate, even if these summary measures reflect ignorable rather than nonignorable nonresponse bias (i.e., the nonresponse bias that can be adjusted using the auxiliary information). There are two key measures at the survey level: the coefficient of variation of subgroup response rates, and its model-based parallel measure, the *R*-indicators (Schouten and Cobben, 2007; Schouten, Cobben, and Bethlehem, 2009). The *R*-indicators, described in more detail in Chapter 9, are functions of the response propensities estimated using the auxiliary information, summarizing the degree of variability in responding in the sample. While they may be critiqued on the degree to which they reflect nonresponse bias in the survey estimates, when used during data collection, they can be informative about the degree of imbalance among the pool of respondents.

Figure 5.2 shows an example of the use of partial *R*-indicators for tracking data collection on the National Survey of College Graduates (NSCG). The NSCG is rather unique as it uses the American Community Survey as a sampling frame, providing a rich set of frame variables—some of which are the same as the key survey estimates, only lagged in time. In 2013, an experiment was conducted in order to increase participation and representativeness in the minority strata (Blacks and Hispanics with bachelor's degrees). The level of effort was increased in the Hispanic and Black strata, and decreased in the White stratum. Based on the partial *R*-indicators over the course of data collection (the lines in the figure) and the weighted response rates (the numbers labeled with 'RR_w' in the right part of the figure), the representativeness in the two minority strata increased in the experimental treatment condition (converged toward zero), compared to the control condition. In this particular instance, however, the effort in the White stratum—the largest sample group—was reduced in such a way that while it contributed to

* Exceptions would be rather atypical, such as the use of the FMI restricted to the inclusion of only one survey variable.

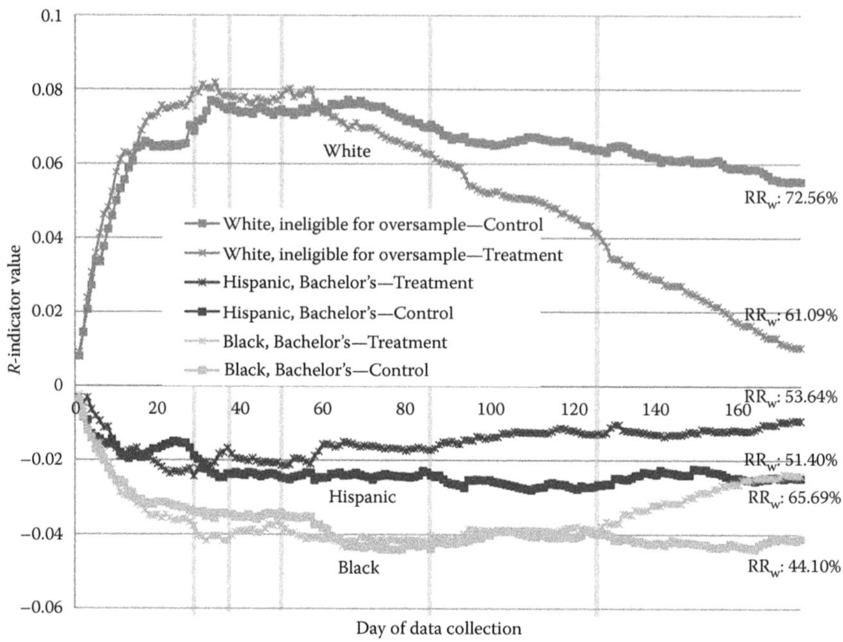

FIGURE 5.2
Unconditional partial R-indicator values and response rates for targeted subgroups in the NSCG, by experimental condition. (Adapted from Reist 2014: https://www2.census.gov/adrm/fesac/2014-06-13_reist.pdf, accessed on April 25, 2017.)

sample balance and a partial *R*-indicator closer to zero, it led to a substantial decrease in the response rate from 73% to 61%.

5.4.2 Estimate-Level Measures

There is tremendous variability in nonresponse bias within surveys (Groves and Peytcheva, 2008). Considering an ongoing data collection, nonresponse bias can be decreasing in one estimate but increasing in another, neither of which may be reflected in a survey-level measure. There are two main types of estimate-level measures: the fraction of missing information (Wagner, 2010), and the association between a survey variable and response propensity (such as means within quintiles or deciles of estimated response propensities).

Rather than reflecting what is known, the FMI, introduced in greater detail in Chapter 9, can be construed as what is left unknown in the sample estimate. The data are multiply-imputed, and the FMI is calculated as the between imputation variance divided by the total variance. If the covariates in the imputation model for a particular variable are not predictive, the FMI will approximate the nonresponse rate. However, if the covariates are highly predictive, the between imputation variance will decrease and the FMI will decline. And here is the key difference between the survey-level measures such as the *R*-indicator and the FMI: while the variation in subgroup response rates indicates ignorable nonresponse bias and penalizes the measure for having highly correlated auxiliary variables, the FMI reflects nonignorable nonresponse bias, after accounting for any highly correlated ancillary variables.

Figure 5.3 shows an example of the use of FMI relative to the nonresponse rate, in NSFG (Wagner, 2010). In this example, the model for 'never married' is quite predictive, in each of the four quarters of data collection, and the FMI is far below the nonresponse rate (although the difference diminishes toward the end of the data collection period).

But this does not mean that the auxiliary variables are just as predictive of other key survey variables. Figure 5.4 shows the same measures, but for 'number of sexual partners in the past 12 months'. Here, the FMI is much closer to the nonresponse rate, and also seems to vary more across quarters. From a monitoring perspective in ASD, one should be more concerned about this variable as it shows greater potential for nonresponse bias.

The survey means within quantiles of the response propensities may be simpler to implement (and automate), but they also share some of the limitations associated with the survey-level measures. If the survey means vary across the propensity quantiles, it can be interpreted as increased risk of

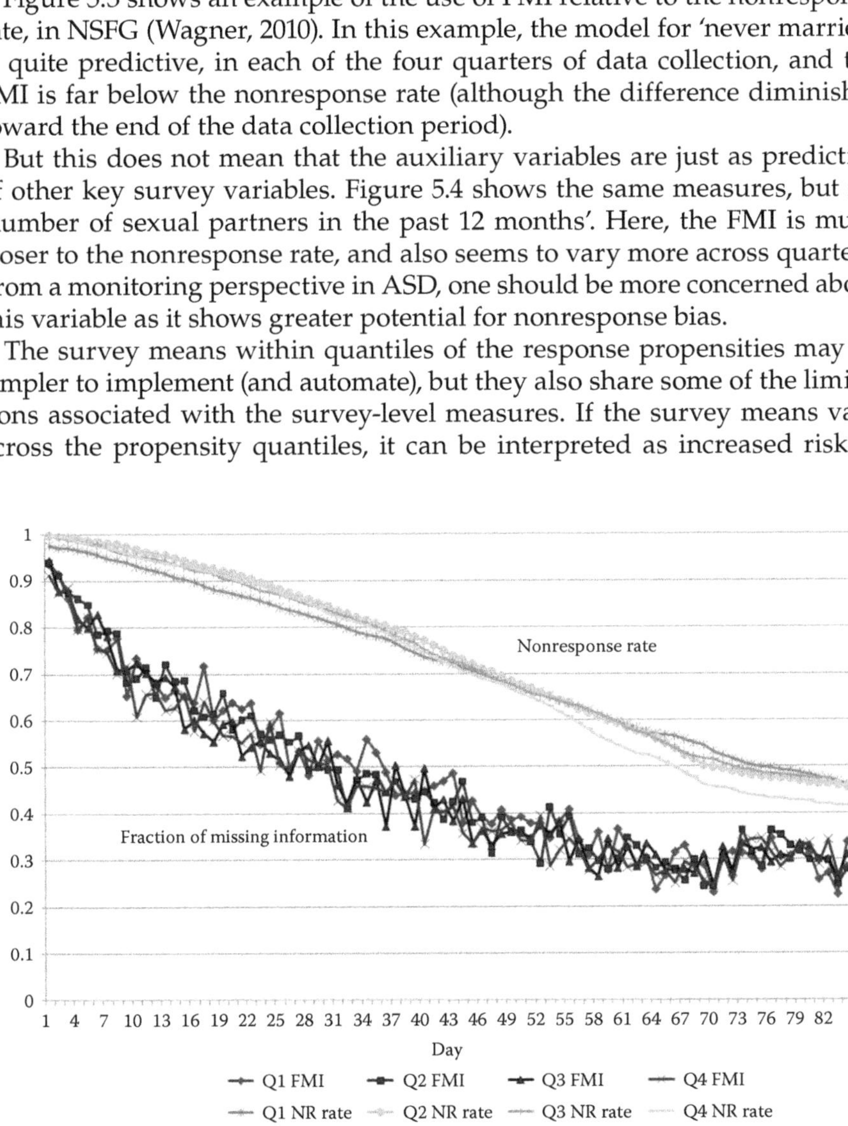

FIGURE 5.3
Fraction of missing information and nonresponse rate for the NSFG by day and quarter: Proportion never married. (Adapted from Wagner, J. The fraction of missing information as a tool for monitoring the quality of survey data. *Public Opinion Quarterly* 74(2); 2010: 223–43.)

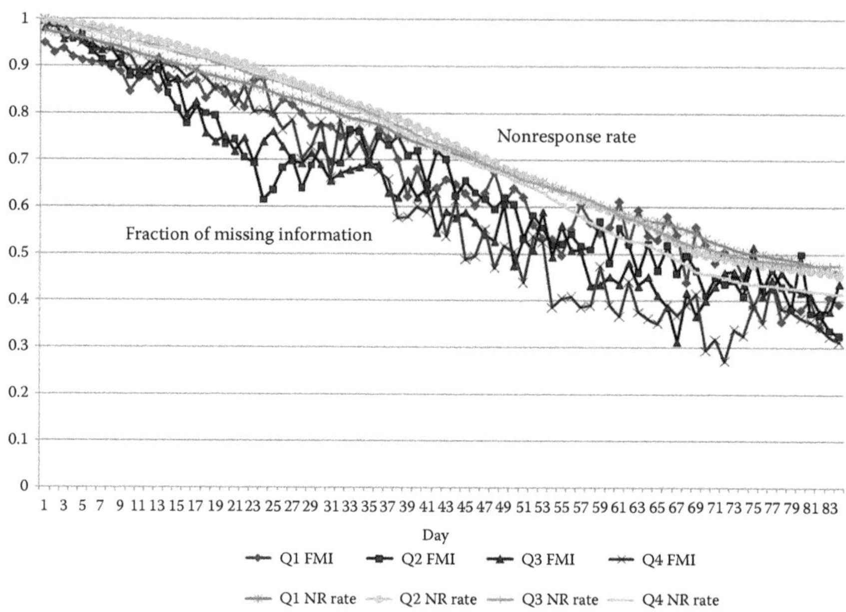

FIGURE 5.4
Fraction of missing information and nonresponse rate for the NSFG by day and quarter: Number of sexual partners in past 12 months. (Adapted from Wagner, J. The fraction of missing information as a tool for monitoring the quality of survey data. *Public Opinion Quarterly* 74(2); 2010: 223–43.)

nonresponse bias in estimates of means based on that variable. However, it presents the ignorable nonresponse bias, assuming that the information that was used to estimate the response propensities will be used in postsurvey adjustments.

5.5 Summary

In this chapter, we introduced models for nonresponse. The first part of the chapter discussed statistical models for directing changes in protocol in an ASD setting, and the relevant characteristics of these models. Key considerations in these models are what data to use, whether to use data from prior or current periods of data collection, and whether to combine past and current estimates.

The second part of the chapter described the main types of measures used to monitor the potential for nonresponse bias during data collection. We divided those in two types, measures at the survey level and measures at the estimate level. Each measure has strengths and weaknesses, and the most appropriate measures in a survey may depend on the amount and type of auxiliary information, as well as the complexity of implementation.

Section III

Implementing an Adaptive Survey Design

6

Costs and Logistics

6.1 Overview

This chapter focuses on two important—if overlooked—aspects of survey design. These are costs and logistics. Accurate cost estimates are essential in order to assess cost-error tradeoffs. Without accurate cost models, it is not possible to identify optimal survey designs. Indeed, without accurate estimates of costs, it may be difficult or even possible to attain the overall objectives of the survey. This is true for adaptive as well as nonadaptive designs. Due to the use of stratification of the population (Chapter 3) and the use of multiple protocols across those strata, ASDs often require very detailed cost models. For example, an adaptive design may need accurate estimates of the costs associated with a Web–telephone sequential mixed-mode design as well as a Web–face-to-face mixed-mode design. Further, these costs may vary across strata due to differences in response rates, interview length, or other factors. In the first part of this chapter, we will examine various cost modeling strategies that may be appropriate for these situations and discuss methods for estimating the parameters of these models. This chapter will use notation, but will carefully elaborate this notation and give examples in order to clarify the meaning. Therefore, these sections are not marked as "technical."

A second topic discussed in this chapter is survey logistics. Logistics involve the methods used for implementing a complex survey. Adaptive survey designs require the development of an infrastructure capable of supporting these designs. These include survey designers and managers familiar with these designs, a management infrastructure for carrying them out, a technical infrastructure for monitoring field progress and triggering the implementation of potentially complex protocol changes, and a method for evaluating and documenting results. In the second part of this chapter, we will look in detail at the infrastructure necessary for the implementation of adaptive designs.

6.2 Costs

An important distinction when considering costs is between fixed and variable costs. Fixed costs are those costs that are independent of the sample size and number of completed interviews. These might include costs aimed at developing the questionnaire, programming the instrument, creating prenotification letters and other materials, designing interviewer training, and so on. Variable costs, on the other hand, are a function of the sample size or the number of interviews. For example, a postpaid incentive is a function of the number of interviews. Other important examples of variable costs include interviewer hours and any materials such as prenotification letters or study brochures that are mailed to sample units. In practice, this distinction may become blurred. For example, the payment of incentives might be more efficiently performed for a large project than for a small project. In general, it is useful to make a distinction between fixed and variable costs.

This distinction is relevant for adaptive designs in at least two ways. First, it may be the case that an adaptive design has higher fixed costs. An adaptive design might use several modes across different strata as opposed to a single mode for the entire sample. This leads to higher fixed costs as each mode requires additional development. Second, the variable costs play an important role in the optimization of the adaptive survey design. Although these variable costs may include many components, usually there are only a few key features that change from one design option to another across the strata. For that reason, it is usually sufficient to focus on a few key drivers when considering a cost model for an adaptive design.

Measuring costs related to surveys can sometimes be difficult. Most surveys operate on fixed budgets and need to have careful cost monitoring. As a result, most data collection organizations have systems in place for monitoring costs. However, these systems may not monitor costs at levels that are relevant for each adaptive survey design. For instance, the costs of administering a telephone survey to a particular stratum in the population may not be measured. Instead, the overall costs of the telephone survey—across all strata in the population—are measured. In the case of face-to-face surveys, the costs of each call attempt are not directly measured. Interviewers report the time at the day level as the number of hours worked each day. The effort applied to each unit of the sample, however, is measured at the call attempt level. This mismatch means that survey organizations may not be able to estimate directly the costs of a particular call attempt or even a particular type of call attempt. As a result, it may be necessary to estimate these costs. Whether these costs parameters are estimated or directly measured, they serve as valuable inputs to cost models that are used to inform adaptive designs.

In this section, several example cost models will be examined. In the first case, a model used by sample designers will be introduced and explained. In the second case, a model used for an adaptive design will be discussed. The goal is to familiarize the reader with the meaning and use of these models. Methods for estimating the parameters in these costs models will also be discussed.

6.2.1 Cost Models

Costs models are not well developed for surveys. The earliest cost models were developed as part of sampling theory (Hansen, Hurwitz, and Madow, 1953; Kish, 1965; Cochran, 1977). These models tended to be quite simple as they were adapted to the requirements of sample design. For example, Kish (1965, p. 268) presents the formula for an optimal cluster size when subsamples are drawn from within clusters in the second stage. The allocation is a function of the unit and cluster costs, as well as a factor known as the intra-cluster correlation. The unit and cluster costs are accounted for in a model similar to the following:

$$C - C_0 = ac_a + abc_b,$$

where C is the total cost, C_0 is the fixed cost, a is the number of clusters, b is the number of units in each cluster, c_a is the cost per cluster and c_b is the cost per unit. The first- and second-stage sample sizes, a and b, are supplied by the sample design. The budget C is provided by the funding agency, the fixed costs C_0 are estimated by the data collection agency. The cost of each cluster c_a and each unit within the clusters c_b is estimated by the data collection agency. This model is used, along with information about the intra-cluster correlation of the survey outcome variable, to determine how large each cluster should be and then solve for the number of clusters that can be afforded given the available budget. The form of this cost model is set by the needs of the sample designer. In practice, costs may not be tracked in this way and it may be difficult to estimate c_a and c_b.

A strength of these early models is that they were stated as mathematical models. That is, the costs could be predicted with estimates of a few parameters. Kish explicitly discusses a general framework for cost models (pp. 263–266). Thus, it seems that more complicated models can fit into this framework. More recent treatments of sample design problems have included parameters to account for nonresponse in various ways. For example, Brick et al. (2002) include parameters associated with screening and main interview response rates across listed and unlisted strata in an RDD sample design. Further, although costs could be treated differently across strata, these cost models did not include differences for different protocols applied to cases in the same sampling stratum. These models express average costs across all cases within the same sampling stratum. For adaptive designs, more detail may be needed.

Groves (1989) is a major exception to the limited discussion of costs in most of the survey literature. He identifies four features of survey costs that do not conform to the simple cost model given earlier. These were discussed briefly in Chapter 2. To review, Groves observes that costs are often nonlinear. For example, the costs of generating the second-stage sample become smaller when overall sample sizes ($n = a \times b$) grow larger. A second problem is that costs are often discontinuous. For instance, there are often "step" functions in costs. One example of a step function is supervision. If one supervisor is needed for every 20 interviewers, when the 21st interviewer is hired, then a second supervisor needs to be added and the costs take a step up. Another step up occurs when the 41st interviewer is hired, and so on. A third problem that Groves describes is that survey costs are stochastic. The estimated costs may be different than the observed costs due to sampling error. The final issue that Groves describes is that survey cost models often have limited domains of applicability. That is, since these models have so few parameters, it may be difficult to generalize them to other settings. In the current setting, this means that a cost model that is appropriate for designing a sample might not work for planning an adaptive design. For example, a cost model suitable for sample design optimization might assume average effort across strata. However, an adaptive design that explicitly differentiates effort across strata might need to include a parameter for effort (e.g., the number of call attempts) so that this effort can be varied across strata.

For these reasons, it is important that cost models be designed with the end purpose in mind. In the case of adaptive designs, this means that the model should include the stratification and design features as parameters. Calinescu et al. (2013) provide an example of developing a cost model for the purpose of optimizing an adaptive survey design. They develop a cost model with inputs for contact rates, refusal rates, different modes, and number of attempts (or "time slots" in their nomenclature). This cost model then allows them to maximize other measures of the data, such as the R-indicator, while maintaining the constraint that this be accomplished within a fixed budget. The model is delineated at the level of the strata to which the differentiated treatment protocols are to be applied (denoted g). This allows each of the parameters (e.g., contact and refusal rates by mode, as well as costs of each by mode) to vary across the strata. For example, if a stratum has a lower contact rate, this will result in the model counting more noncontact cases and the costs associated with obtaining a noncontact and relatively fewer interviews and the costs associated with these interviews.

This approach has great flexibility. The cost model proposed by Calinescu et al. (2013) is a useful example that merits a detailed explanation since it is explicitly built for an adaptive design. Let $\mathcal{g} = \{1,\ldots,G)$ be strata in the population. These units within the strata are assumed to be homogenous with respect to the impact of the design features on their probability of responding. We note that it is also useful that they be homogenous with respect to the survey outcomes as well (see Chapter 3). This would make it more likely that an adaptive design could have an impact on estimates. Defining these strata allows us to calculate separate costs for each. From the point of view of the design, it allows us to estimate other design parameters (e.g., contact rates by mode) separately for each stratum. These estimates can differ. This is essential for an adaptive design as it allows us to identify the differential protocols that we will want to apply across the strata. Let $\mathcal{T} = \{1,\ldots,T\}$ be the "time slots." In an interviewer-mediated survey, these "slots" may represent contact attempts. In self-administered surveys, these slots can be times at which the sampled unit is approached with the aim of soliciting a response to the survey. The modes are then denoted $\mathcal{M} = \{1,\ldots,M\}$.

They also estimate contact probabilities $p_g(t,m)$, where t and m denote that these estimates are for a particular time slot and mode while g denotes that these estimates can vary across strata in the population. For example, $p_1(3,Web)$ would be the estimated probability of contact for the third time slot when approached via the Web for stratum 1 in the population. Similarly, there are also response probabilities $r_g(t,m)$, where t and m denote that these estimates are for a particular time slot and mode while g denotes that these estimates can vary across strata in the population.

Finally, they denote the survey costs using $b^s(m)$ to denote the costs of a successful completion of an interview in mode m. There are also costs estimated for failed attempts. Since these costs might differ depending upon the outcome, they define $b^{fc}(m)$ as the costs for a failed contact attempt in mode m. The cost of a failed attempt at obtaining an interview after achieving contact is denoted $b^{fr}(m)$. With this notation, it is possible to define the average cost of attempts to a particular group in the following manner:

$$p_g(t,m)\left[r_g(t,m)b^s(m)+(1-r_g(t,m))b^{fr}(m)\right]+\left(1-p_g(t,m)\right)b^{fc}(m). \tag{6.1}$$

Equation 6.1 can be broken down into components in order to be more easily understood. The first part of the equation, $p_g(t,m)[r_g(t,m)b^s(m) + (1 - r_g(t,m))b^{fr}(m)]$ is the cost of completing an interview in time slot t in mode m. In brackets are the costs of successful interview attempts times the rate of success of those attempts $r_g(t,m)b^s(m)$ plus the costs of failed interview attempts times the rate of failure (i.e., 1 minus the rate of success) of those attempts $(1 - r_g(t,m))b^{fr}(m)$. These costs are then associated with the proportion of the sample that is contacted via the contact rate in front of the brackets $p_g(t,m)$. The second part of the equation $(1-p_g(t,m))b^{fc}(m)$ is the costs of the failed contact attempts. It includes the rate of noncontact (one minus the rate of contact) and the costs of a noncontact. In this model, the costs change depending upon the mode assigned to the group at each time slot.

Table 6.1 shows some hypothetical example rows using this cost model. The table does not include all strata or time slots. For reasons of brevity, it only includes two strata and two time slots. The probabilities of contact $p_g(t,m)$ and interview $r_g(t,m)$ come from Calinescu's et al. (2013) paper, but the

TABLE 6.1

Hypothetical Example Cost Model Calculations Based upon Calinescu et al. (2013)

Stratum (g)	Time (t)	Mode (m)	$p_g(t,m)$	$r_g(t,m)$	$b^s(m)$	$b^{fr}(m)$	$b^{fc}(m)$	Average Cost per Attempt	Average Cost per Interview
1	1	FtF	0.3	0.9	5	4	2	2.87	€10.63
	2	FtF	0.4	0.7	5	4	2	3.08	€11.00
1	1	Tel	0.4	0.8	3	2	1	1.72	€5.38
	2	Tel	0.5	0.5	3	2	1	1.75	€7.00
2	1	FtF	0.8	0.9	5	4	2	4.32	€6.00
	2	FtF	0.6	0.7	5	4	2	3.62	€8.62
2	1	Tel	0.7	0.8	3	2	1	2.26	€4.04
	2	Tel	0.6	0.6	3	2	1	1.96	€5.44

cost parameters $\{b^s(m), b^{fc}(m), b^{fc}(m)\}$ are made up in order to illustrate how the model works. This table could be expanded with more strata, time periods, and modes. Focusing on just the first row, it contains the average cost per attempts given the contact and interview rates and the expected associated costs for cases in stratum 1 at the first attempt (i.e., time slot), for each of these types of outcomes (limited to interview, refusal, and no contact): $0.3*[0.9*5 + (1 - 0.9)*4] + (1 - 0.3)*2 = 2.87$. It is possible to estimate the number of interviews per attempt as the product of $p_g(t,m)$ and $r_g(t,m)$. For example, in stratum 1 at time 1 using the face-to-face mode, the probability of an interview for each contact attempt is $0.3*0.9 = 0.27$. We can rebase the costs per attempt as costs per interview by dividing the average cost per attempt by the probability of an interview at each attempt. This is in the last column of the table.

It is possible to make a more detailed model. For example, more outcome categories could be included in the model. The three used in this model may not capture enough of the variation in the lengths (i.e., cost) of different attempts. For example, attempts with contact but no interview may be another important category that takes more time than a failed contact attempt or refusal attempt, but less than an interview. If this category of outcome also occurs at different rates across the strata, then it may be important to include it in order to capture the cost differences across the strata.

In general, the complexity of the cost model is a function of the design. The model needs to include all the strata and design features that will be varied. Then important parameters, such as the contact and response rate in the example, also need to be included. Including all the strata allows the effectiveness of the same protocol to vary over strata in the population. For example, in Table 6.1, the contact rates for the same attempt ($t = 1, t = 2, \ldots$) vary across the strata. The contact rate in stratum 1 for attempt 1 via the telephone is 0.4 while it is 0.7 for the same attempt in stratum 2. This complexity is necessary in order to identify designs that are more effective or less costly for some strata. If the effectiveness or costs of the design do not vary across the strata, then the one-size-fits-all design is the best design. Second, it allows for the realistic assumption that costs may vary depending upon response rates. For example, an unsuccessful contact attempt can have a different cost from a successful contact attempt that results in a completed interview. We note that failed contact attempts are assumed to have the same cost across strata. However, this assumption, while reasonable in this case, is not necessary. It is possible to test this assumption with existing data, if available. If the assumption is not reasonable, then the same outcome type (e.g., interview, refusal and no contact) could have a different cost estimate across the strata.

In the example put forth by Calinescu et al., (2013) the design feature that is varied is the modes that are deployed and the timing of those modes. As an example, Calinescu et al. (2013) developed an optimal solution for one of the strata in their simulated data (p. 119). The solution is given in Table 6.2. While the example in Table 6.1 includes two time slots, this table shows all six time slots. The solution indicates that for this stratum, the first attempt in the first time slot should be a face-to-face attempt, the second time slot should be a telephone attempt, the third time slot should not have an attempt, the fourth time slot should be another telephone attempt, and the final two time slots should not have attempts.

Given estimates of the contact rate, participation rate, and costs of failed contact attempts, failed attempts to complete the interview with those contacted, completed interviews at each time and for each mode, and a total sample size for stratum $g = 2$, it is possible to calculate the costs for the solution in Table 6.2 following the example in Table 6.1. Alternatively, the cost model can be used to determine whether a specified set of protocols (one for each group), given the size of each group, will lead to going over a specified budget limit. In this way, the costs of many possible designs can be measured against a specified total budget. This budget is a constraint on a resource allocation problem. Then, optimal solutions (i.e., allocating the various modes to the time slots for each group) can be identified that maximize some quantity (e.g., the response rate or the R-indicator) without going over the specified budget limit. This is exactly the approach Calinescu et al. (2013) take. Problems of this type are known as resource allocation problems (see Bather, 2000 for a more detailed discussion). These sorts of problems are part of a larger field of study known as operations research. Optimization is discussed in more detail in Chapter 7.

TABLE 6.2

Optimal Solution for $g = 2$

Time Slot (t)	**1**	**2**	**3**	**4**	**5**	**6**
Mode (m)	F2F	Ph	0	Ph	0	0

6.2.2 Cost Model Parameter Estimation

Cost models require input parameters that may be more or less detailed depending upon the complexity of the problem. Estimating the input parameters for these models can frequently be difficult. In this section, we discuss several strategies for estimating these parameters.

There are several reasons that explain the difficulty of estimating survey cost model parameters. First, costs may not be tracked at the appropriate level of detail. For example, interviewers may report their hours worked on a daily basis and not at a call attempt level. Therefore, there is no record of how long successful and unsuccessful call attempts take, or any kind of call attempts for that matter. Second, cost information may be seen as proprietary across survey organizations. Releasing this information may lead to a competitive disadvantage. Therefore, while information can be shared within organizations, it is often the case that cost information about surveys is not published or otherwise made available to other survey organizations. Third, costs may not be linear in the sample size, so that costs in one survey design cannot be translated directly to another design when sample sizes change drastically.

There are several strategies for dealing with these issues. Two example strategies are suggested. The first is to make use of paradata to estimate cost parameters. In a computer-assisted telephone interviewing (CATI) setting, it may be possible to measure the length of a call from the paradata, which may include timestamps. If the call attempt is timestamped at the beginning and end of the call, then the length of the call can be directly calculated. If the call attempt is only timestamped at the beginning of the call, then the length of the call can be estimated by the amount of time before the next call by the same interviewer is begun. This method has some issues (e.g., the length of the last call made on any shift cannot be estimated, interviewer breaks will be mistakenly added to the call attempt preceding the break, etc.), but it can produce reasonable estimates.

In an in-person study, an estimate derived from timestamps in the call records may be much less reliable. In this case, the paradata might be used in a statistical model that estimates the cost parameters. For example, it is often possible to combine call record data with timesheet data in order to estimate call lengths. The simplest model is to divide the total hours worked by the total number of calls. This gives an average call length. In some settings, this simple estimate may suffice. However, for models such as the one detailed earlier, this type of estimate is not sufficient. It may be necessary to estimate more complicated models. Recall that survey organizations often ask interviewers to report their hours worked at a daily or shift level. From the call record data, we can also count the numbers and types of calls made on each of those days. Given this data structure, it is possible to use regression to estimate how long each type of call takes. Table 6.3 contains fictional data from interviewer days. The data include the hours worked each day and the number of calls made that resulted in contact and the number that did not result in contact.

TABLE 6.3

Fictional Call Summary and Hours Data for Four Interviewer Days

Interviewer	Day	Contact Calls	No Contact Calls	Hours
1	1	4	6	3
1	2	1	0	0.4
2	1	0	3	0.75
2	2	8	2	4

If we estimate a regression model that predicts the number of hours worked using the number of contact calls and the number of no contact calls, then the coefficients from that model are estimates of how long each type of call takes. Table 6.4 shows the estimated coefficients for the two types of calls from the data in Table 6.3.

The results are estimates that contact calls take about 0.44 hours (26.4 minutes) while no contact calls require 0.22 (13.2 minutes). These estimates could be used in a cost model similar to the one described earlier.

This technique can be used in many settings. Many more types of calls can be included. For some surveys, it may make sense to include an intercept term. This intercept term represents the hours spent preparing to interview. In surveys with travel, it may be necessary to include parameters for this travel. For example, between cluster travel might be captured by including a count of the number of clusters visited on each day. In some cases, the length of an interview is known. This might be the case in a CAPI setting where the beginning and end of the interview are timestamped. If these times are known, they do not need to be estimated. These known times can be subtracted from the hours worked. In that case, the estimated coefficient for completed interviews might represent the amount of time it takes to talk to the respondent before and after the interview.

Table 6.5 is an example of the estimates from a more detailed model. The model is predicting the hours worked in a week by an interviewer for a survey with two modes—face-to-face and telephone. Each record in the dataset represents an interviewer week. It includes the hours worked by the interviewer that week, and a count of the calls of each type made that week. The table presents the estimated coefficients. These are also converted to time (hours and minutes, e.g., 4.62 hours equals 4 hours and 37 minutes).

TABLE 6.4

Estimated Coefficients from a Model Predicting Hours Worked

	Estimate	Standard Error	t-Value	$Pr(>\|t\|)$
Contact Calls	0.44	0.01	33.03	0.00
No Contact Calls	0.22	0.02	12.59	0.01

TABLE 6.5

Estimated Coefficients for a Regression Model Predicting Hours Worked per Week

Mode	Variable	Estimated Coefficient	Hours:Minutes
	Intercept	4.62	4:37
Face-to-Face	No contact	0.14	0:08
	Contact	0.27	0:16
	Set appt	0.32	0:19
	Resistance	0.84	0:50
	Final noninterview	1.51	1:30
	Final nonsample	3.78	3:46
	Interview	1.48	1:28
Telephone	No contact	0.12	0:07
	Contact	0.26	0:15
	Set appt	0.28	0:16
	Resistance	0.33	0:19
	Final noninterview	0.99	0:59
	Final nonsample	0.87	0:52
	Interview	1.12	1:07

This model includes an intercept. The intercept represents time that may be required each week to carry out administrative tasks such as attending meetings. These tasks are not associated with the number of calls each week. The types of call are then separated by mode. The expectation is that similar call outcomes will take a different length of time in each of the two modes. The estimates bear that out as the telephone attempts generally take less time than the face-to-face attempts of a similar type. The estimates indicate a face-to-face "no contact" call takes 0.27 hours or 16 minutes. An interview conducted over the telephone takes 1.12 hours or 1 hour and 7 minutes. These estimates can then be used in complex cost models, such as those described earlier in this chapter. Further, these estimates have standard errors that may be useful when conducting a sensitivity analysis (see Chapter 8 on robustness).

In face-to-face surveys, travel costs are an important component of the total budget. They may constitute 20%–40% of the entire budget. Travel costs usually have two components (1) the time the interviewer spent in travel, and (2) travel-related expenses which are usually expressed as a particular rate per mile traveled using their own automobile but can also be reimbursement for bus, train, or airplane travel and even hotel charges for longer trips. However, travel costs may be difficult to estimate. It is difficult to assign travel costs to a single case. Since travel is usually to clusters of cases, these costs are "shared" by several cases. This makes it difficult to assign a specific travel cost to any case. If a change in cluster size is being contemplated, then travel costs can be conceived of as the number of visits required to each cluster in order to complete the cluster. This approach requires being able to estimate the varying number of trips as a function of the sample size. If that can be done, then the problem of assigning costs to specific cases is avoided. When interviewers are assigned samples from multiple projects, then the problem of assigning travel costs to specific cases is complicated. If the travel costs are apportioned across the projects, then they depend on the mix of projects an interviewer is assigned to at any moment.

Another strategy for estimating the parameters from cost models is to conduct a special study. It might be possible to use an expensive measurement technique to measure costs on a small sample—either a small portion of the full sample of a survey, or a sample of the population of surveys. Kalsbeek et al. (1994), for example, used field staff to follow National Health Interview Survey interviewers and record how long each of their calls took. The survey is conducted face-to-face. This special study allowed them to estimate the lengths of different types of calls. These estimates were then used to derive call limits.

Another type of special study involves the use of GPS tracking. These data can be used to record interviewer movements. These data are complex, but may be used to understand how long different calls take, how much time is spent traveling between clusters, and so on (Olson and Wagner, 2015).

Finally, expert opinion can often be used to create estimates. This is often the source of creating survey budgets for a survey for which there is little or no prior experience. Expert opinion might include discussing with survey managers, careful review of existing information about related studies, and talking to staff that carry out the tasks being estimated. Of course, data are always preferred, but the opinion of experienced survey professionals is often quite accurate and may serve as a useful starting point.

6.3 Logistics

Planning and implementing ASD is a complex activity. In order to be successful, ASDs require specialized procedures for designing and planning the survey, management infrastructure for implementing the design, a technical infrastructure for monitoring the survey, and methods for executing specified interventions, such as mode changes or controlling the number of contact attempts. Finally, ASDs should be evaluated and documented so that the experience and lessons learned can be the basis of improved designs in the future.

It may be useful to start slow and grow—that is, to start from relatively simple designs, build upon existing infrastructure, and add features iteratively. For example, it may be that existing structures can be used as the basis for adaptive designs. Many CATI sample management systems will allow for the use of multiple languages, and match sampled units who speak a particular language to interviewers who also speak that language. Perhaps this feature could be used to assign a particular calling protocol to a stratum. Assigning the sample predicted to have the lowest probability of responding to interviewers with the highest ratings could be accomplished using this feature. It may also be the case that "compromise" procedures will be needed. For example, although an adaptive design may call for switching from telephone to face-to-face interviewing after two telephone calls, it may be that this switch will be handled only at specified points in time by data managers. This may results in some cases receiving more telephone calls than desired while other cases receive fewer than desired. Nevertheless, these first steps will be important as they may spur the development of improved infrastructure.

It is also useful to distinguish between real-time adaptive interventions or "responsive design" from preplanned adaptive survey design, or "static" adaptive design. The former requires detailed monitoring systems and the ability to implement interventions, which requires having staff trained to do the monitoring and make decisions and who have the technical ability to push these decisions out to interviewers. For instance, this might require the ability to change the mode assignment of cases meeting particular criteria. Static adaptive designs also require trained analysts. However, the requirements of the interventions can be spelled out before the field period and the systems designed to carry out the prespecified adaptations.

This section discusses the management and technical infrastructure elements that may be required to implement ASDs. While these designs are relatively new, organizational experience is growing. As these designs evolve, infrastructures will also evolve.

6.3.1 Stages of Implementing Adaptive Survey Designs

The first step in an adaptive survey design is planning. Before beginning with the design of the survey, it is important to identify the most important objectives. As was discussed in Chapter 2, different surveys can place different values on controlling costs, limiting biases, or reducing variance. These priorities can then be used to inform the inexactly defined tradeoffs between costs and errors that most surveys face.

Once the priorities have been defined, the next step for an ASD is to identify the key risks. What are the major sources of error and most important cost drivers? Kirgis and Lepkowski (2013) describe how the NSFG team used paradata to identify key challenges faced by the NSFG Cycle 6. They used these lessons to redesign the NSFG 2006–2010. In many cases, surveys are facing the challenge of reducing their budgets without reducing the quality of the data they collect. In these circumstances, it seems natural to look to the use of inexpensive modes of data collection, such as the Web. However, these less expensive modes may lead to increased nonresponse and measurement bias.

Once the risks have been identified, the design team can begin to identify important strata in the population and design protocols that can meet the objectives of the survey. In the case of surveys that are attempting to reduce budgets without significant harm to the quality of estimates, survey designers may seek to limit measurement biases by assigning sampled persons to modes in which they are less likely to give incorrect responses to questions. Further, they may seek the mixture of modes for recruitment that produces the most balance with respect to observed covariates on the sampling frame as a way to limit the risk of nonresponse bias. This is just the type of problem addressed by Calinescu et al. (2013).

Identifying recruitment strategies across multiple strata can be a difficult task. Of course, this task requires staff who can analyze data from previous surveys. In Chapter 4, we discussed a potential set of interventions and targeted strategies. There are two sides of the problem. These need to be addressed in tandem. The first is to identify a set of strategies or interventions that will achieve the targets (e.g., staying within a fixed budget, maximizing R-indicator, and minimizing risk of measurement error). The second is to identify subgroups for which differential strategies may either come at a reduced cost or minimize some error source. For example, Chesnut (2013), in an analysis of American Community Survey (ACS) data, suggested that there is a subgroup in the population for which the probability of responding via the Web is very near zero. The ACS uses a Web–mail–telephone–face-to-face sequential mixed-mode design. Chestnut's suggestion is that for some, skipping the Web mode might be beneficial. The costs of attempting this mode can be saved for this group. It may also reduce their annoyance so that later modes would be more effective.

Once the plan has been developed, the survey may require special infrastructure in order to be implemented. On the management side, adaptive survey design requires managers familiar with the techniques. This can be accomplished via a well-designed training program. Chapters 2, 3, 4, 7, and 9 of this book might form the core of such a program. In the end, experience is the best teacher. It may take time for this experience to accrue. One approach is to start with relatively simple designs with two or three strata and a small number of variable design features. More strata and features can be added as experience accumulates.

6.3.2 Monitoring

An important aspect of the management of adaptive designs is a technical infrastructure capable of monitoring the survey process and intervening in that process. The monitoring component includes the ability to receive updated information in a timely manner. This information can be presented as a "dashboard." There are several key criteria for designing the content of such a dashboard. The first criterion is relevance—that is, is the proposed figure, table, or indicator relevant to the quality or costs of the adaptive design? Does it inform a decision related to these elements? The second criterion is timeliness. The dashboard needs the information to be available at a frequent enough basis that it can be used to make decisions. If some data are not available until late in the field period (e.g., travel expenses), then proxy measures that are available on a more timely basis should be identified. The third criterion is that the dashboard should allow users to visualize uncertainty. This is related to the discussion of robustness in Chapter 8. In the context of adaptive designs, visualization of uncertainty will aid managers avoid making premature decisions based on scant evidence.

In addition to meeting these criteria, the dashboard should include critical measures in each of the following areas:

1. Survey inputs (e.g., interviewer hours)
2. Efficiency (e.g., measures of interviews per hour)
3. Survey outputs (e.g., number of interviews) and
4. Survey quality (e.g., response rates, subgroup response rates and changes in key estimates)

This will allow managers to make key decisions or implement prespecified design changes in a timely manner.

In practice, it is impossible to design a dashboard that will answer every question. The opposite extreme—no prespecified analyses—is just as untenable. This approach is represented by Figure 6.1a below. In this type of organization, where there is no prespecified analyses, the analysts can become a bottleneck. This may reduce the amount of monitoring that is done.

On the other hand, it is not feasible to monitor surveys without some preplanned and, if possible, automated analyses of paradata. Thus the key indicators can be routinely monitored, while analysts are available to answer

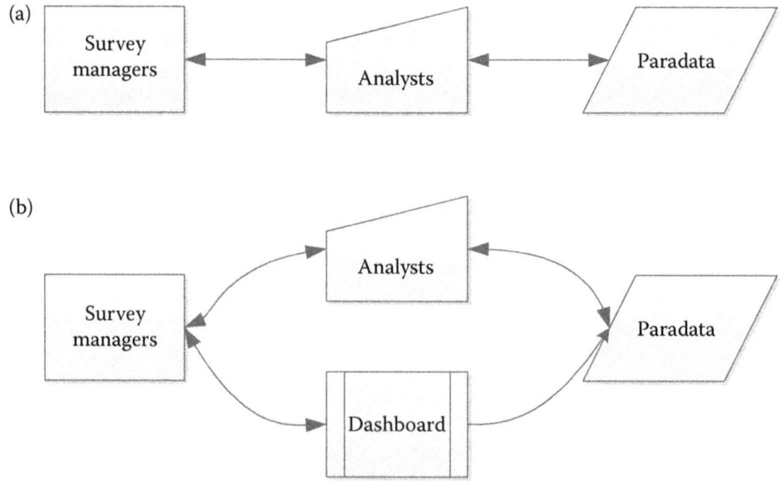

FIGURE 6.1
Organizational models for turning paradata into information for managers.

unanticipated questions or aid in the design of new surveys. This is the approach outlined in Figure 6.1b.

Developing these technical systems can be a challenge. There is a dearth of survey literature on dashboard construction for surveys. There are, however, more general discussions of how to build and design dashboards (e.g., see Few, 2006) that may be helpful. An example of one approach to building dashboards for a survey project is given at the end of this chapter.

In addition to information flows, adaptive design also requires the technical capacity to implement design features which may be implemented at the level of the sample unit. An adaptive design might require switching modes based on the outcome of a prior time period or action. For example, this could require implementing a rule such as "if there is no response one week after an email invitation to participate in the survey has been sent, then the case will be made available for telephone interviewing." A protocol such as this, with decision rules defined at the level of the sample unit, requires a sample management system capable of implementing a complicated set of rules. The system has to be aware that the email invitation was sent and that a week has passed without an interview being completed. Then it must transfer the sample unit to a CATI interviewer. These actions must be implemented repeatedly throughout the field data collection across a large number of sample units. As mentioned earlier, it may be possible to implement these kinds of rules manually with some "slippage," and allow the development of automated systems later.

One of the simplest kinds of interventions has been to prioritize cases. The bases for these prioritizations are discussed in Chapter 4. Here, we discuss the technical means for implementing these measures. In telephone studies, case prioritization is governed by the call scheduling algorithm. These algorithms may be as simple as sorting the sample into the order in which it should be called. In this case, prioritization can be accomplished by moving the prioritized cases to the top of the sorted list. In other cases, this might involve more detailed procedures that attempt to place prioritized calls at times during the day when the highest average contact rates occur. Even more elaborate schemes would attempt to identify the times when each particular household is most likely to be contacted.

In face-to-face surveys, case prioritization can take on a more rudimentary form. In these surveys, interviewers are often largely able to determine their own schedules and decide when to call cases. In this setting, prioritization can take the form of communicating a greater importance of some cases to the interviewer so that they can place more calls or make calls at the best times for the prioritized cases. The technical systems need to allow the communication of which cases are prioritized. This can be done using a simple "flag" in the interviewer's view of the sample management system. Figure 6.2 shows an example of such a flag. In this case, the prioritized cases are identified by exclamation points in the column labeled "Priority."

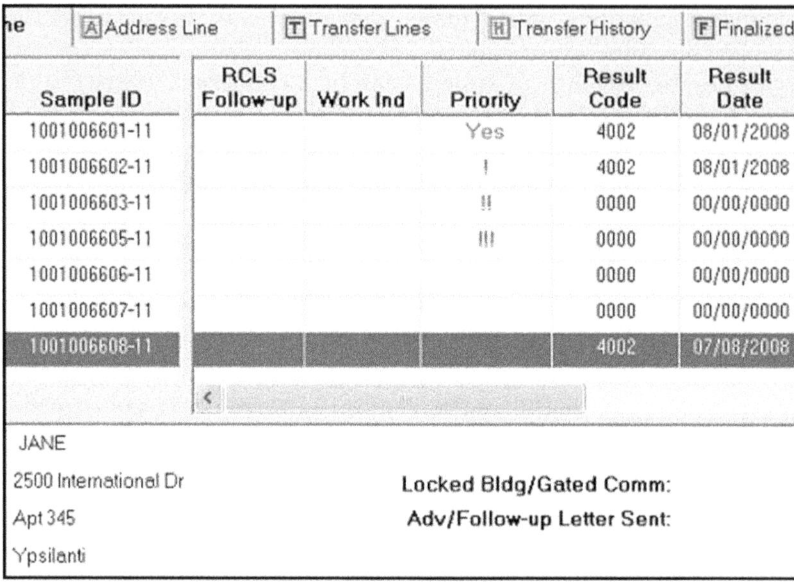

FIGURE 6.2
Case prioritization in a face-to-face survey as seen in the sample management system.

Development of the management and technical infrastructure for adaptive survey designs will be an evolutionary process. A useful approach may be to start with relatively simple adaptive designs, and develop the technical and managerial infrastructure required for it. Then this design can be elaborated in future surveys. As the design is elaborated, so are the technical and managerial means necessary for the implementation of new features.

6.4 Examples

6.4.1 The Dutch LFS—Continued

The different (multi-mode) strategies in the LFS vary greatly in their expected costs. In Chapter 4, nine strategies were described that have been tested for the LFS: W, TS, TE, FS, FE, W → TS, W → TE, W → FS and W → FE. Here, "W," "T," and "F" are short for Web, telephone, and face-to-face, and "S" and "E" are abbreviations for standard and extended fieldwork effort. The "→" refers to a sequential strategy where nonrespondents to Web receive a follow-up. The stratification for the LFS was discussed in Chapter 3, and has nine strata based on age, size of the household, registered unemployment, registered employment, and ethnicity.

Table 6.6 presents the estimated expected relative costs per sample unit for each of the eight modes and standard or extended fieldwork effort strategies, and for each of the strata. The extended face-to-face strategy is set to one and functions as a benchmark. The costs include all variable costs, but exclude fixed costs associated with overhead and design of questionnaires and other material. As expected the face-to-face strategies are most expensive and the Web strategies are the cheapest. The sequential strategies have costs that are in between those of the single-mode strategies. The costs differ across strata as well because they differ in their average interview duration, their average number of required reminders, calls and contacts and their differences in telephone coverage and response propensities. However, these differences are not as pronounced.

LFS logistics faced two main challenges: the case management over survey modes and the linkage of auxiliary variables. The first challenge, the case management systems, arose from three factors: the timing and planning of the survey modes, the variability in the Web response rates, and the face-to-face interviewer fixed contracts. The latter contracts specify a number of hours that interviewers must be paid for working each week. The face-to-face interviewers are mostly employees with fixed hour contracts, and, hence, with fixed workloads, while the telephone interviewers work on flexible contracts. The employee contracts for the face-to-face interviewers originate from the desire to maintain a well-trained and consistent team of interviewers. In order to reach efficiency in travel costs, the face-to-face interviewers receive geographically clustered sets of sample addresses. As a result, for the case management, some amount of time is needed to identify a geographically clustered set of nonresponding Web cases and reallocate them to a face-to-face interviewer. One week is needed for telephone but face-to-face requires several weeks due to the fixed workload and the need to cluster addresses of nonresponding

TABLE 6.6

Estimated Expected Costs Per Sample Unit per Strategy and Stratum

	Stratum								
	1	2	3	4	5	6	7	8	9
W	0.03	0.04	0.04	0.04	0.04	0.03	0.03	0.03	0.03
TS	0.11	0.15	0.10	0.09	0.13	0.11	0.09	0.12	0.14
TE	0.13	0.17	0.11	0.10	0.15	0.14	0.11	0.16	0.19
FS	0.84	0.89	0.83	0.83	0.86	0.84	0.81	0.84	0.89
FE	1.00	1.00	1.00	1.00	1.00	1.00	1.00	1.00	1.00
W→TS	0.08	0.11	0.09	0.09	0.09	0.08	0.08	0.07	0.07
W→TE	0.09	0.12	0.10	0.10	0.10	0.09	0.09	0.08	0.07
W→FS	0.60	0.66	0.64	0.70	0.59	0.56	0.65	0.51	0.61
W→FE	0.72	0.71	0.80	0.84	0.73	0.68	0.81	0.62	0.71

Note: The costs for the extended single mode face-to-face strategy (FE) are set to one.

Web cases. However, as a side-effect, the time lags between Web and the other modes do offer the advantage to adapt strategies to population strata, if the case management systems were to allow for a differentiation. For this purpose, the case management systems had to be changed. It was not expected, but the response rates in the Web mode fluctuate strongly from one month to another, and led to problems in managing the interviewer workloads. In some months, when the Web response rates were lower than expected, the face-to-face workloads were too large for available interviewing staff. In other months, when the Web response rates were higher than expected, the workloads were too small. This problem was solved by subsampling Web nonresponse to fixed workloads per month. The subsampling rates were set such that even with the lowest realized Web response rates, the sample size could achieve some minimum. This fixed sample size in the second mode allows for stratification in order to again adapt to the population strata.

The second challenge in logistics, the linkage of auxiliary variables, is the result of varying degrees of timeliness and accuracy of the administrative data. The auxiliary variables from administrative data play an important role in stratifying the population before and during data collection. The more accurate and timely the auxiliary variables are, the more informative and effective the interventions can be. However, for legal reasons, administrative data are physically separated from case management systems. Furthermore, since samples are drawn from municipality registers, some amount of time is needed to check, and possibly update, address and demographic information from sampled persons or households through municipality authorities. Finally, some amount of time is needed to load a sample into the case management systems and to link telephone numbers from commercial databases. Consequently, the data collection preparation requires little over a month in total, so that administrative data needs are linked 6 weeks in advance of data collection. Since auxiliary variables may become outdated during data collection, an intermediary update needed to be planned and performed. The linkage of auxiliary variables led to an additional process that runs parallel to data collection.

6.4.2 NSFG—Continued

In a recent experiment, the NSFG needed to estimate the costs for two different designs, one with a US$40 phase one incentive, and the other with a US$60 phase one incentive. Both arms received an US$80 incentive in phase two (see Wagner et al., 2017, for a more detailed description). The interviewers had assignments including samples from both treatments. Interviewer hours were reported at the day level, and not separately for each treatment group. Therefore, it was not possible to directly calculate the hours of effort expended on each treatment group. Instead, two approaches were employed to estimate the costs associated with each incentive treatment. The first used a simple average cost per call. The other method used the regression approach described earlier in this chapter.

The approach based on the average length of time per call attempts assumes that each call takes about the same amount of time. We know that this assumption is not true. The most readily evident contrast is that interviews take longer than noncontacts. The average length of a call was estimated by dividing the total number of interviewer hours by the total number of call attempts. Using this approach, a call is estimated to take about 25 minutes on average. The higher incentive led to a reduction of 2.5 calls per interview. This yields an estimated savings of about 62.5 minutes per interview. In this case, the estimate obtained using the average length of all call attempts is likely an overestimate of the savings as the attempts that were saved were shorter than average (i.e., fewer noncontacts and fewer repeated visits).

A more complex model regresses the hours an interviewer worked each week on the number of calls of different types (interviews, no one at home, setting appointments, etc.) that the interviewer made that week, to estimate the lengths of time different types of calls might take. The coefficients from such a model are presented in Table 6.7. The predictors include an intercept, the number of days worked in a week, and the number of areas segments visited. The predictor gives a sense of administrative and travel costs. Interviewers typically work 30 hours per week and spend 15%–30% of their time in these activities. The number of days worked further helps define travel costs as each day an interviewer works they need to travel from their home to sampled area segments. The number of segments visited gives a measure of how much travel there is between area segments. The estimated coefficient indicates that each trip to a new segment adds about 45 minutes. The other estimates are for different types of call attempts. One of the coefficients is negative (main contact). This seems to indicate that the intercept and other coefficients actually include time for call attempts, but that when contact with an identified eligible person is made, this increases the efficiency of that week. Estimates of the hours for each incentive treatment from this

TABLE 6.7

Estimated Coefficients from a Model Predicting Hours Worked in a Week

Predictor	Estimated Coefficient	Standard Error
Intercept	7.85	0.31
Number of days worked	1.64	0.10
Main interviews	1.38	0.04
Screening interviews	0.24	0.02
Main contact	−0.07	0.02
Screening contact	0.11	0.03
Main no contact	0.15	0.01
Screening no contact	0.06	0.00
Nonsample	0.25	0.03
Number of segments visited	0.75	0.05

regression model lead to predicted savings of about 48 minutes per interview under the US$60 treatment. This estimate is based on the observed number of call attempts of each type that are made for each of the incentive treatments. The US$60 treatment led to increased efficiency by reducing the number of calls in specific categories relative to the US$40 treatment.

An interviewer hour in the NSFG (excluding recruitment, hiring, and training costs, and taking salary and nonsalary costs into account) costs about US$35. At this rate, the additional incentive cost of US$20 is expected to produce savings in interviewer costs between US$28.00 (48 minutes savings estimated from the regression model) and about US$36.50 (the 62.5 minute estimate using the average length of a call method) per interview. There are some other savings that occur in phase two as a result of the higher incentive in phase one. Overall, the net savings per interview are between US$10.00 and US$19.00 after factoring in the additional cost of the higher incentive and assuming that the higher incentive has the same administrative cost structure as the lower incentive. These estimates are based on experimental data. The savings might be different in practice, since the experiment adds some administrative costs associated with tracking the assigned treatment over different cases. Overall, the cost savings were small, but the higher incentive did cover its own cost with some increased efficiency for interviewer effort.

An additional issue faced by the NSFG, was the development of a dashboard tool for monitoring field progress. Kirgis and Lepkowski (2013) give a description of how this was accomplished for NSFG 2006–2010. Data are extracted from the sample management system, timesheets, and the questionnaire data on a daily basis using SAS. These data are processed into a summarized format in SAS before being exported to Excel. These summaries include information at the day level and the interviewer level. Comparison summary information from previous quarters and years are not updated daily. As a quarter is completed, the daily- and interviewer-level data are summarized and exported to Excel for that quarter and year in order to serve as comparisons. In Excel, figures were built based upon the summarized data. Figure 6.3 is an example of such a figure. The figure shows the ratio of screener to main calls by day. This ratio is important since the project seeks to focus on screening effort in the first few weeks. In addition to the current quarter's daily ratio and 7-day moving average, the figure shows two comparison lines that represent the ratio from previous quarters or the average of several previous quarters.

The figures are organized in a "table of contents"' that includes categories such as "Effort," "Active Sample," "Productivity," and "Data Set Balance." The first two categories are measures of inputs, while the latter two categories are measures of output. "Effort" includes figures such as interviewer hours, counts of call attempts, and the ratio of main call attempts to screener call attempts. "Active Sample" includes figures that characterize the amount and quality of the active sample available to interviewers. It includes figures such as percentage of the sample not finalized, number of lines not attempted, and percentage of cases with any resistance. "Productivity" measures interviews

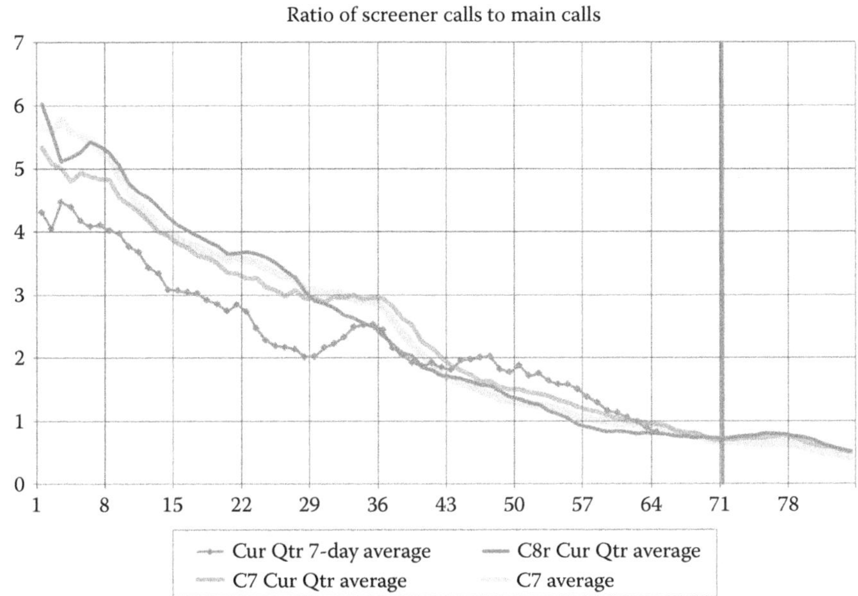

Figure 6.3 Example figure from the NSFG dashboard.

and the efficiency of those outputs—the total number of main interviews completed, the total number of screening interviews completed, and the hours of interviewer effort expended per completed main interview are examples. "Data Set Balance," measures the quality of the interviewing and includes indicators such as response rate, balance with respect to sampling frame data, interviewer observations, and screening data such as the presence of children in the household.

In recent years, a new dashboard has been built that is based upon keystroke data from the interview. The aim is to identify issues that occur at the interviewer level in the administration of the questionnaire. For example, does the interviewer tend to administer items more quickly than other interviewers? Does the interviewer back up more frequently than other interviewers? These measures are calculated weekly and converted to z-scores. Then supervisors contact any interviewer that is more than three standard errors from the mean of these statistics for 3 or more weeks in a row. They discuss the issue and identify strategies for improving the administration of the questionnaire.

6.5 Summary

For ASDs, cost models that are based upon the key design features that are to be adapted and the strata across which these design features will vary should be included in the model. This chapter gave an example of such a cost model, and discussed methods for estimating the parameters in this model.

Understanding survey costs are a difficult endeavor. Some costs are measured directly and can be accurately estimated. In many situations, however, information gathered as paradata such as call records are not directly linked to cost information such as interviewer hours and expenses. This means that at least some survey costs will have to be estimated using some sort of model. These models can be validated by using them across multiple projects. The goal is to provide accurate estimates so that optimal designs can be identified. Further, social changes such as the rising rates of nonresponse also impact costs and create an environment of uncertainty. In practice, estimates are likely to be at least somewhat inaccurate. In Chapter 8, we review the implications of these inaccuracies for the overall efficiency of adaptive designs.

Improving cost estimates, either through more careful elaboration of statistical models to estimate key parameters, special studies to measure key costs, or both of these approaches are necessary to improve all survey designs, including adaptive designs.

Logistically, adaptive designs also require specialized management and technical infrastructures. The development of these structures may proceed in parallel with the elaboration of adaptive designs. That is, as new adaptive features are developed, the technical and managerial means for implementing these designs may be developed. These initial steps may lead to the development of fully elaborated systems that are quite different than those that are currently in use. This is difficult to predict.

This chapter described how existing technologies (e.g., SAS and Excel) were harnessed to create flexible monitoring systems. In the face of uncertainty about what the future may hold, starting with small scale, easy-to-use systems and elaborating these over time should help to avoid the problem of deploying large scale, expensive systems that are suited for survey designs that are no longer used.

7

Optimization of Adaptive Survey Design

7.1 Introduction

So far we have seen ASD as a non-iterative process in which we start with the main objectives in the survey, identify the likely threats to achieving the desired outcomes, specify alternative protocols to vary during data collection or identify possible changes to the main protocol, develop models to tailor the design to different sample members, and assign protocols to sample members. While the ASD allows for tailoring the protocol to sample members, we are still making a deterministic presumption that we know what ASD to use. Instead, here we consider the initial ASD as our best attempt to improve the study design and implementation, and that there are a number of ways that the initial ASD can be improved.

Consider the Interactive Case Management (ICM) adaptive design in the National Intimate Partner and Sexual Violence Survey (NISVS) introduced in Chapter 2 in which paradata are used to predict the likelihood that a sample telephone number would yield an interview. In this approach, logistic regression models are estimated in order to identify sample telephone numbers with extremely low response propensities. Such sample cases can then be put on hold, allowing interviewers to spend their time on the more productive cases. If only we stop cases that almost never produce interviews, then we could expect to also increase response rates by letting interviewers spend the time they would have been dialing these unproductive numbers on numbers that may yield additional interviews. There are two incredibly important decisions in this design: *when* and *which*.

When do we implement the ICM intervention? In a 12-week data collection we can implement it as early as after the first week, as late as after the 11th week, or any time in between (and even multiple times during the data collection period). If we fit the models too early in data collection, we will likely fail to identify the unproductive sample numbers, as the models are informed by the outcomes of the call attempts to each sample number. If the number has been called only once that single outcome would likely not provide a prediction of the final outcome at the end of the survey with sufficient precision. That is, a number that was dialed once and nobody answered may or may not yield an interview by the end of the study. But a number that has been dialed five times over the course of several weeks and none of the calls have been answered, may be predicted fairly accurately by the model as unlikely to yield an interview.

Conversely, we could implement the intervention in week 11—toward the end of data collection—when predictions become accurate. However, at that point there is little potential benefit from the intervention, as the cost savings from not dialing unproductive sample telephone numbers and reallocation of interviewing effort will be limited to only the last few days of data collection. Therefore, our goal is to implement the intervention as soon as the model-based predictions become sufficiently accurate.

There are also multiple ways of operationalizing *when*. We could say that the intervention will be implemented after 4 weeks of data collection. But if a telephone number was dialed only twice, we may not want to trust the model. Thus, we may also want to have a minimum number of call attempts to a sample case before it is eligible for the intervention. Similarly to the timing during data collection (i.e., week in data collection), the additional unknown is the minimum number of call attempts with which we could have sufficiently accurate estimates of whether the sample number would yield an interview.

Which sample telephone numbers should be stopped? The model predictions can sort the sample numbers by the likelihood of yielding an interview, but we could stop as few as the lowest 1% or as many as 99% of the sample cases, or use any propensity threshold in between. If the threshold is very low, there could be very minimal impact from the intervention, if any. However, if the threshold is set very high, potential interviews may be missed that could lower response rates.

So while the ASD is clearly specified in this example—use a propensity model on past data and apply the coefficients to the current data collection outcomes in order to identify which sample cases to put on hold—the parameters for *when* to implement and *which* cases to stop are unknown quantities that can be varied. That is, the design can be further optimized. Just as adaptive designs address uncertainty in data collection outcomes, there is inherent uncertainty in the adaptive designs themselves. In this chapter, we present the main approaches to optimization of ASDs and we use the ICM example to illustrate each approach. Although we use this example for illustration, we also add examples from other studies that have different designs. In particular, many surveys have auxiliary information from the sampling frame, external databases, and prior waves of administration, and are not restricted to paradata generated after the start of data collection.

There can be considerable uncertainty in any optimization, which is not the focus in this chapter; Chapter 8 addresses the factors that contribute to uncertainty in ASD and presents methods for sensitivity analysis.

7.2 Approaches for ASD Optimization

A distinction can be made between two main approaches to optimization, which can be used in combination: (1) numerical optimization, which includes mathematical and statistical methods, and simulation on existing data and (2) trial and error. When optimizing a limited set of features of the survey in isolation of all other features, a *mathematical optimization* may be possible. In this approach, an assumption is made that many design features do not interact. For example, one may be optimizing the allocation of effort across samples from two overlapping sampling frames with respect to cost and measurement error, yet the standard formulae will ignore any changes in the impact of the dual-frame estimator (how the data from the two samples are combined). Similarly, a design may be optimized for balancing measurement error with nonresponse bias, assuming that the two error sources do not interact with other features such as the level of effort, interviewer pool, etc.

Thus, a related alternative that can also be used in conjunction with mathematical methods is *simulation on existing data*. In this approach, different scenarios are simulated on prior survey data and evaluated based on the study's objectives. In the prior example, it could involve the creation of multiple datasets that satisfy the cost constraints, and for each scenario, weight the data and compute the desired outcomes such as precision and bias in survey estimates.

Even if a mathematical or simulation-based solution is found, it is still conditional on the parameters used in the optimization. Some may be derived from prior data (thus conditional on a particular design and subjected to changes over time), while others may be assumptions based on best judgement and similar studies. A *trial and error* approach can be desirable not only when relevant prior data are not available, but also when different features of the design can interact in unknown ways. For example, consider an optimization of the number of interviewers to employ on a survey. A larger number of interviewers can reduce interviewer variance as it is a function of the number of interviewers, but it generally increases cost, as training and supervision costs increase. Given cost constraints, one could optimize the number of interviewers to increase the precision of the survey estimates. Yet, this optimization would ignore other impacts of the change that are not observed in the data—for example, the intraclass correlation coefficient reflecting how similar responses are within interviewers has been found to be larger in studies that use more interviewers with smaller workload. Thus, without an experiment, the optimum number of interviews based on a formula will likely be suboptimal.

There are other distinctions that can be made between optimization methods. A second classification is whether ASDs try to exactly fulfill budget or other constraints such as the exact response rate or number of respondents (limit incidental variation), or try to do so only on average over multiple runs of the survey (limit structural variation).

A third distinction is whether ASDs lean on empirically quantified interaction between actions/design choices and sample strata or that they rely on softer, more qualitative experience based choices. For example, consider an optimization of the mix of mail and telephone modes on a survey. Another survey on a different topic, conducted using a different sampling design and on another population and with similar but not identical data collection procedures may help inform the ASD on a new survey—but the information may be treated as more qualitative due to the differences in designs. At an extreme, qualitative information collected from debriefing interviews with the interviewers could be used to tailor procedures in an ASD, which may lead to trial and error optimization, discussed in more detail later in this chapter.

A fourth distinction is whether optimization needs to be performed during data collection because there is no prior data (one-time surveys) versus optimization in between waves. In general, optimization in surveys with replication (repeated cross-sectional surveys, surveys with continuous data collection, longitudinal surveys, and surveys with rotating panel designs) have more data and more opportunities to optimize their ASD; having past data collected under the same or similar design provides prior parameter estimates for models, data for simulations, and empirical evidence from experimental tests. However, a one-off survey may involve the greatest uncertainty in cost, response rates, achieved sample size, and other objectives, increasing the potential benefit from an ASD.

7.3 Numerical Optimization Problems

7.3.1 Mathematical and Statistical Optimization

One approach is to find a mathematical optimum for multiple objectives, within cost constraints (Calinescu, 2013; Schouten, Calinescu, and Luiten, 2013). In this approach, objectives are parameterized as functions, such as various response quality functions (e.g., response rate, *R*-indicator, coefficient of variation, and estimated nonresponse bias), and a cost function (which includes both fixed and variable costs). Schouten et al. (2015) show how these multiple objectives can be set up as nonlinear optimization problems that can be handled by standard statistical packages, although they also note that there are specialized optimization software packages that can handle the computation more efficiently and with fewer estimation problems.

In their example, they show how interviewers that vary in their ability to gain cooperation can be assigned to sample members who have been previously interviewed, and for whom response propensities can be calculated.

The optimization was used to identify which sample members to assign to which group of interviewers in order to maximize the *R*-indicator, within cost/interviewing constraints.

A major strength of this approach is the ability to optimize the design for more than one outcome, or to include multiple constraints. For example, we have implemented similar approaches in practice, identifying which sample members to assign to the more experienced interviewers at the beginning of the study (Peytchev, 2010), but this is done only in settings where sample members are assigned to multiple telephone interviewers and the assignment only applies to the first contact. That is, in order to make the assignments for the duration of the data collection, further constraints will need to be added, beyond the estimation of response propensities.

A weakness of this approach, relative to the other optimization alternatives, is the need for strong assumptions about parameters in the model (e.g., response rates for experienced and less experienced interviewers) and that the estimated/assumed cost and error functions will not fall apart when a change is implemented. To elaborate on the latter, consider a cost model that is based on 40 interviewers, each working 20 hours per week. The model may be reasonable for estimating the cost of 10 additional interviewers or if the workload is increased by 5 hours, but it may fall apart if the number of interviewers is doubled or the interviewers are asked to work full-time at 40 hours per week, as the cost structure can change—supervisors may need to be hired and different employment benefits may need to be provided. For more detailed discussion on this, see Groves (1989). In practice, however, the quality and cost objective functions are relatively smooth, so that small deviations in input parameters do not have a strong impact on realized cost and quality. Nonlinear mathematical optimization problems can be used to address such problems, but are inherently hard to solve, often necessitating the use of local optima under the possibility of more optimal solutions.

Another limitation of this approach is the amount of complexity that can be included. For instance, in the example from Schouten et al. (2015), what if experienced interviewers require fewer contact attempts in order to finalize a sample case if the sample person is reluctant, compared to less experienced interviewers? This is one of many interaction effects that can occur, but is near impossible to specify all such possibilities. These interactions hamper any form of optimization, whether adaptive or not. A key implication is the need for modesty in adaptation, avoiding drastic changes to the design that have not been evaluated empirically.

A related challenge is that as many parameters need to be estimated, one very soon reaches the boundaries of precision. In studies of mode effects at Statistics Netherlands, the estimated mode effects were found to have very wide confidence intervals within strata. This lack of precision can make it very hard to decide what action is best, unless very large pilot/experimental studies are done first. However, it is important to estimate the level of imprecision in order to avoid decisions based on such estimates—see Chapter 8 for a relevant discussion on sensitivity analysis.

As another example of this limitation, consider a common problem in telephone surveys in which the allocation between landline sample numbers and cell phone sample numbers is optimized for cost and precision. A typical approach is to use a formula for optimum allocation in dual-frame telephone surveys (Brick et al., 2011a) based on general dual-frame optimization (Hartley, 1962). The inputs include the size of each population (landline and cell phone) and the cost per interview from each sampling frame. However, there are many other factors that are ignored primarily for simplicity, such as the different selection methods in each frame that results in different loss in precision due to weighting. In a landline interview, a common design is to select only one adult in the household, and this adult receives a weight equal to the inverse of the number of eligible adults in the household. The allocation optimization formula, however, assumes that the variability in the selection weights across the two sampling frames is equal. The next method addresses some of these limitations.

7.3.2 Simulation on Existing Data

Simulations do not require for all relevant factors to be explicitly parameterized, and can be a useful complementary analysis to validate the mathematical and statistical methods. In simulations using survey data, a set of features is modified and applied in the form of "what-if" scenarios.

Consider the last example with a dual-frame telephone survey. We noted that we could use a formula for a mathematical optimization of the sample allocation between landline and cell phone numbers, but that the formula cannot account for complexities such as the impact of weighting under different allocation scenarios, as the impact of each weighting adjustment (selection within household, number of landlines, number of cell phones, nonresponse adjustments, etc.) is nearly impossible to include. A much simpler approach would be to simulate different sample allocations on prior data, and for each simulated allocation, to fully weight the data and compute bias and variance estimates (Peytchev and Neely, 2013).

Let us return to the ICM example we presented at the beginning of the chapter, in which we posed the problem of identifying unproductive sample cases during data collection in a telephone survey. Recall that to optimize the ASD, we are uncertain when to stop any cases (defined by day in data collection and minimum number of call attempts) and how many cases to stop (what propensity threshold to use).

If we were to use only statistical methods without simulation, we would pick a "reasonable" point in data collection to implement these and some minimum number of call attempts for cases to be eligible. We could then estimate a logistic regression model estimating the likelihood of a case to become an interview on prior data (or alternatively, current data collection) and apply the regression coefficients to the current data collection to estimate the propensity of yielding an interview. We would then select some threshold, possibly based on the distribution of estimated propensities and desired number of cases to stop, and stop calling to cases that fall below that threshold.

Given prior data, however, we could simulate "what-if" scenarios by varying the day in data collection, the minimum number of call attempts, and the propensity threshold (Peytchev, 2014). We could then evaluate for different scenarios what proportion of call attempts would be avoided (cost saving or time made available for more productive sample cases) and what proportion of interviews would be missed (the opportunity cost of stopping low propensity cases). Table 7.1 shows the proportion of calls that would be avoided, by day and minimum number of call attempts, if a propensity threshold of $p = 0.0005$ is used. For example, if the ICM is implemented on the 76th day of data collection (10 days before the end of data collection) for sample cases with a minimum of 13 call attempts, 1.4% of the call attempts would be avoided.

Table 7.2 shows the potential cost. For the same combination (76th day, at least 13 call attempts, and $p = 0.0005$ threshold), 0.2% of the interviews are expected to be missed due to the implementation of the adaptive

TABLE 7.1

Daily Proportion of Calls Still Made, by Implementation Day and Minimum Number of Call Attempts, $p = 0.0005$

Day	Call 7	Call 8	Call 9	Call 10	Call 11	Call 12	Call 13	Call 14	Call 15	Call 16	Call 17	Call 18
68	0.987	0.987	0.987	0.987	0.987	0.987	0.987	0.988	0.988	0.989	0.993	0.995
69	0.989	0.989	0.989	0.989	0.989	0.989	0.989	0.990	0.990	0.991	0.992	0.994
70	0.988	0.988	0.988	0.988	0.989	0.989	0.989	0.990	0.990	0.991	0.992	0.993
71	0.987	0.987	0.987	0.987	0.987	0.988	0.988	0.988	0.989	0.990	0.992	0.994
72	0.983	0.983	0.983	0.983	0.983	0.983	0.983	0.984	0.984	0.986	0.988	0.990
73	0.985	0.985	0.985	0.985	0.985	0.985	0.985	0.985	0.986	0.987	0.989	0.991
74	0.985	0.985	0.985	0.985	0.986	0.986	0.986	0.986	0.987	0.988	0.990	0.992
75	0.982	0.982	0.982	0.982	0.982	0.982	0.983	0.983	0.983	0.984	0.986	0.988
76	0.985	0.985	0.985	0.986	0.986	0.986	0.986	0.986	0.986	0.988	0.990	0.991
77	0.992	0.992	0.992	0.992	0.992	0.992	0.992	0.992	0.993	0.993	0.994	0.995
78	0.990	0.990	0.990	0.990	0.990	0.990	0.990	0.990	0.991	0.991	0.992	0.993
79	0.990	0.990	0.990	0.990	0.990	0.990	0.991	0.991	0.991	0.992	0.993	0.994
80	0.990	0.990	0.990	0.990	0.990	0.990	0.990	0.990	0.990	0.991	0.992	0.993
81	0.983	0.983	0.983	0.983	0.983	0.983	0.983	0.983	0.984	0.984	0.985	0.987
82	0.991	0.991	0.992	0.992	0.992	0.992	0.992	0.992	0.992	0.992	0.993	0.994
83	0.987	0.987	0.987	0.987	0.987	0.987	0.987	0.987	0.987	0.988	0.989	0.990
84	0.989	0.989	0.989	0.989	0.989	0.990	0.990	0.990	0.990	0.990	0.992	0.992
85	0.978	0.978	0.978	0.978	0.978	0.979	0.979	0.980	0.980	0.981	0.983	0.984

Source: Adapted from Peytchev, A. Models and interventions in adaptive and responsive survey designs, DC-AAPOR Panel on Adaptive Survey Design. Washington, DC., February, 2014. Accessed March 26. Available at http://dc-aapor.org/ModelsInterventionsPeytchev.pdf.

TABLE 7.2

Cumulative Proportion of Completed Interviews, by Implementation Day and Minimum Number of Call Attempts, $p = 0.0005$

Day	Call 7	Call 8	Call 9	Call 10	Call 11	Call 12	Call 13	Call 14	Call 15	Call 16	Call 17	Call 18
68	0.999	0.999	0.999	0.999	0.999	0.999	0.999	0.999	0.999	0.999	0.999	1.000
69	0.999	0.999	0.999	0.999	0.999	0.999	0.999	0.999	0.999	0.999	0.999	1.000
70	0.999	0.999	0.999	0.999	0.999	0.999	0.999	0.999	0.999	0.999	0.999	1.000
71	0.998	0.998	0.998	0.998	0.999	0.999	0.999	0.999	0.999	0.999	0.999	1.000
72	0.998	0.998	0.998	0.998	0.998	0.998	0.998	0.998	0.998	0.999	0.999	0.999
73	0.998	0.998	0.998	0.998	0.998	0.998	0.998	0.998	0.998	0.998	0.999	0.999
74	0.998	0.998	0.998	0.998	0.998	0.998	0.998	0.998	0.998	0.998	0.999	0.999
75	0.998	0.998	0.998	0.998	0.998	0.998	0.998	0.998	0.998	0.998	0.999	0.999
76	0.997	0.998	0.998	0.998	0.998	0.998	0.998	0.998	0.998	0.998	0.999	0.999
77	0.997	0.997	0.997	0.997	0.997	0.998	0.998	0.998	0.998	0.998	0.999	0.999
78	0.997	0.997	0.997	0.997	0.997	0.997	0.998	0.998	0.998	0.998	0.999	0.999
79	0.997	0.997	0.997	0.997	0.997	0.997	0.998	0.998	0.998	0.998	0.999	0.999
80	0.997	0.997	0.997	0.997	0.997	0.997	0.997	0.998	0.998	0.998	0.999	0.999
81	0.997	0.997	0.997	0.997	0.997	0.997	0.997	0.998	0.998	0.998	0.999	0.999
82	0.997	0.997	0.997	0.997	0.997	0.997	0.997	0.997	0.998	0.998	0.998	0.999
83	0.997	0.997	0.997	0.997	0.997	0.997	0.997	0.997	0.997	0.998	0.998	0.999
84	0.997	0.997	0.997	0.997	0.997	0.997	0.997	0.997	0.997	0.998	0.998	0.998
85	0.997	0.997	0.997	0.997	0.997	0.997	0.997	0.997	0.997	0.998	0.998	0.998

Source: Adapted from Peytchev, A. Models and interventions in adaptive and responsive survey designs, DC-AAPOR Panel on Adaptive Survey Design. Washington, DC., February, 2014. Accessed March 26. Available at http://dc-aapor.org/ModelsInterventionsPeytchev.pdf.

design—only about 2 in 1,000 interviews would not be completed under this adaptive design, because they were completed after the 76th day, had at least 13 call attempts, and were among the sample cases identified by the model as having a response propensity below 0.0005. That is, under this very cautious scenario for an initial test of the ASD, a modest saving in unproductive call attempts would be made, but virtually no interviews would be missed. These scenarios were then run for different propensity thresholds.

An even more sophisticated simulation is to determine when to stop the study altogether. From a nonresponse bias perspective, a study has reached its end when the additional completed interviews do not change the estimates. In the presence of rich auxiliary information, this can be treated as a bias-prediction problem, imputing the data for all the nonrespondents. As soon as the estimates based on the fully-imputed data are not statistically different from the observed data, the study can be stopped as continuing with the same protocol is unlikely to change the estimates (Wagner and Raghunathan, 2010).

Apart from optimizing the adaptive design on a survey, the simulation approach can also be used to evaluate whether an adaptive design would even be beneficial. Särndal and Lundquist (2014b) show that reliance on weighting for nonresponse can be reduced. Schouten et al. (2015) used data from several surveys to simulate that focusing effort to balance the sample through an adaptive design could be superior to reliance on postsurvey adjustments. It is worth noting that under certain conditions sample allocation in adaptive design also warrants changes to weighting, which is discussed in Chapter 10.

The simulation approach shares some limitations with the mathematical optimization as it relies on assumptions in order to extrapolate beyond the observed data (collected under the original protocol). In the example in which sample cases were being stopped based on a very low likelihood of yielding a completed interview, we can simulate how many interviews would be lost if effort is truncated to those telephone numbers. However, we cannot simulate how many additional interviews could be achieved resulting from interviewers spending more time calling more productive sample cases, as those outcomes are not observed in the prior data used for the simulation.

7.4 Trial and Error

The trial and error approach addresses the main limitations of the mathematical and simulation-based optimization methods by varying features in the ASD, implementing these variations, and evaluating them for future implementation. This approach is particularly well-suited for repeated cross-sectional surveys (e.g., biennial) and ongoing surveys with multiple sample releases (continuous surveys with monthly or quarterly data collection), as they allow for experimentation that informs subsequent implementations of the survey—and gradual optimization of the ASD.

A key component of the trial and error approach is randomization—using an experimental design—for direct evaluation of the adaptive design on a survey. In the example from NISVS that we have been using in this chapter, the ICM approach was tested experimentally toward the end of the last data collection period in 2013. Sample telephone numbers were randomly assigned to the treatment or the control group, in each sample (landline and cell phone). The models (for landline and cell phone samples) were estimated approximately 10 days before the end of data collection and sample numbers in the treatment condition that fell below a propensity threshold were stopped. The threshold was more liberal for the landline sample where the model also had shown better fit (see Chapter 8 on imprecision in models used in ASD); 46.1% (4,897 cases) of active landline cases and 5.7% (2,032 cases) of active cell phone cases were stopped. We then evaluated the cost (in interviews) and benefit (in number of call attempts) from the ASD intervention.

Figure 7.1 shows the number of interviews in the treatment and in the control conditions, by sample type. The difference between the two conditions

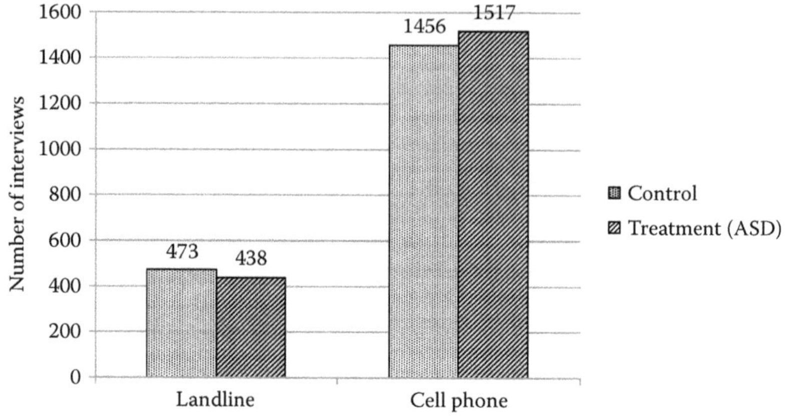

FIGURE 7.1
Number of interviews by control and treatment (ASD) condition, in the landline and cell phone samples. (Adapted from Peytchev, A. Models and interventions in adaptive and responsive survey designs, DC-AAPOR Panel on Adaptive Survey Design. Washington, DC., February, 2014. Accessed March 26. Available at http://dc-aapor.org/ModelsInterventionsPeytchev.pdf.)

is not significant in either of the samples. So even though a large proportion of the cases were stopped in the landline sample, it did not affect the number of completed interviews. Of course, this result is conditional on when the ASD intervention was implemented—10 days before the end of a 12-week data collection period.

Figure 7.2 shows the benefit from this ASD. The average number of call attempts in the treatment condition is significantly lower than in the control

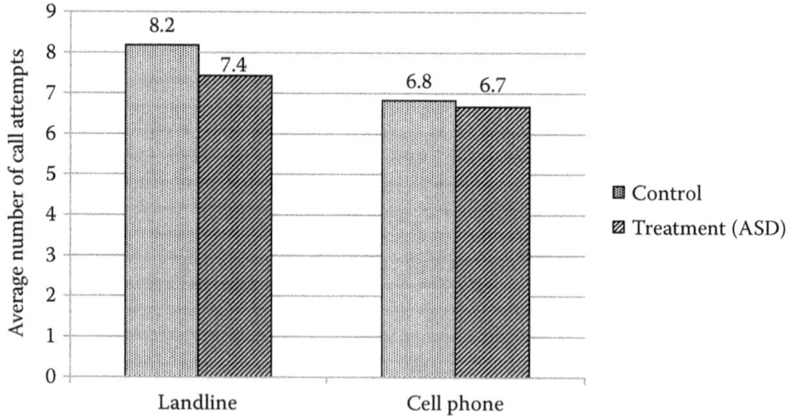

FIGURE 7.2
Average number of call attempts by control and treatment (ASD) condition, in the landline and cell phone samples. (Adapted from Peytchev, A. Models and interventions in adaptive and responsive survey designs, DC-AAPOR Panel on Adaptive Survey Design. Washington, DC., February, 2014. Accessed March 26. Available at http://dc-aapor.org/ModelsInterventionsPeytchev.pdf.)

condition, in both landline and cell phone samples. It is not surprising that the difference is larger in the landline sample, where almost half of the cases were stopped, while only about 1 in 17 cases in the cell phone sample was stopped. These differences in means may not seem large, but multiplied by thousands of cases, result in preventing thousands of unproductive call attempts—substantial amount of interviewer time that could be spent on dialing more productive sample numbers or other aspects of the survey.

Trial and error may also be used when applying a developed ASD to another survey, and more importantly, the randomization can become a stable feature of the design for ongoing improvement. The California Health Interview Survey (CHIS), a state-wide RDD telephone survey with continuous data collection, implemented the ICM approach in 2015. The simulation that was presented in Tables 7.1 and 7.2 using NISVS data was repeated using data from the first quarter of data collection. In order to implement this adaptive design feature with caution, the threshold for stopping cases was initially set very low in the second quarter and gradually increased/adjusted with each subsequent quarterly sample release. Here as well, cases were randomly assigned to control and treatment conditions in order to evaluate the impact of the ASD. To increase the benefit of the treatment (ASD) condition, the proportion of the sample assigned to the treatment condition was also increased in later sample releases.

The examples so far have been from surveys with little, if any, auxiliary information available prior to data collection. Many surveys benefit from rich sampling frames. Panel surveys also generate data on sample members that can be used in the following waves of data collection. Rich auxiliary information, particularly variables that are associated with survey measures, provide a range of opportunities, including ASD models to identify which sample cases to prioritize and what treatments to assign. The National Center for Education Statistics conducts several large national surveys, repeated every few years. The National Postsecondary Student Aid Study (NPSAS) examines the characteristics of students in postsecondary education, with a focus on how they finance their education, and is conducted every 4 years. The sampling frame starts with schools from which student lists are obtained. As part of the study, auxiliary data are collected on the institutions and on the students, including data from student aid applications, loans, and grants that contain key survey variables. Some of these data become available prior to data collection, others during data collection, and have been used to model response propensities and related measures (i.e., Mahalanobis distances) to identify cases and groups of sample cases that are underrepresented among the respondents. The adaptive design consists of identifying sample cases that may be most likely to contribute to nonresponse bias, and using more intensive protocols on these cases. This process can be repeated at several points (phases) during data collection. Models and interventions are devised prior to data collection and trained on prior data, but an essential component is trial and error. In the NPSAS and its related studies, it takes four forms.

First, a pretest is fielded, which is of sufficient size to test a limited number of experimental conditions. Here, different interventions can be tested, along with identifying any interactions with characteristics of sample members. The results then inform the ASD used on the main study.

Second, the main study is evaluated for whether the ASD had an impact on the survey estimates, both bias and variance. Based on these evaluations, decisions are made whether to test other approaches on the next data collection.

Third, several surveys use the NPSAS as baseline measures and as the sampling frame for their own data collection, such as the Beginning Postsecondary Students (BPS) longitudinal study and the Baccalaureate and Beyond (B&B) longitudinal study. These surveys, along with other NCES-sponsored studies such as the High School Longitudinal Study (HSLS) share many data collection characteristics and are coordinated together by a team of researchers at NCES and RTI International (conducting the data collection). This provides the opportunity for these studies to use variations of the same general ASD and learn from each other's experiences (e.g., Cominole et al., 2013; Pratt et al., 2013).

Fourth, trial and error optimization was used on the BPS study by testing procedures on an initial ("calibration") sample release (10% of the full sample), to inform the procedures for the remainder of the sample. While this is exemplary of responsive design, in this instance this phased design is used to determine the optimal intervention for the ASD. The experiment manipulated the effect of different incentive amounts (or "dosage" of the treatment, as described in Chapter 4), from US$0 to US$50 in increments of US$5, and evaluated the incentive effect on response rates for five groups defined by response propensity. Figure 7.3 shows the results from this calibration of the treatment after 6 weeks of data collection. The increases in response rates were significant for each additional US$5 from US$0 to US$30. This information was used to decide on using US$30 for the remaining (90%) of the sample.

Responsive designs tend to rely on trial and error and show how the same approach can be used in ASDs. For example, what incentive amount should be used in the last phase of data collection for sample members who have remained nonrespondents to that point? Numerous incentive amount tests have been conducted on surveys, but the applicability of these results may be very limited due to differences in study designs. The same question was faced on the NSFG, where two different incentive amounts were tested during data collection. Based on the incoming results, a decision was made to continue the survey with the higher incentive amount.

A key strength of this approach is the ability to improve an ASD when prior data are not available, or may simply lack relevance to the current survey design. Prior studies may provide an indication of what may be a preferable set of design features, but without sufficient certainty or ability to evaluate tradeoffs. In the NSFG example, one could assume that a higher incentive

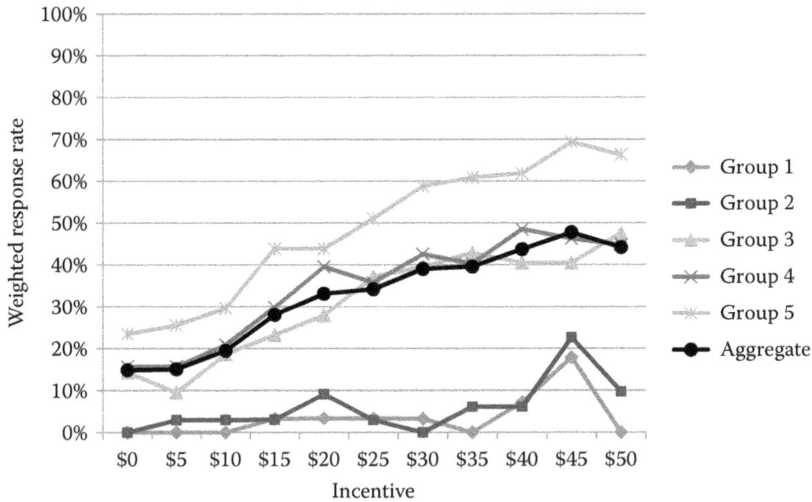

FIGURE 7.3
Weighted response rates by incentive amount and response propensity group in BPS 2014. (Adapted from Wilson, D., B. Shepherd, and J. Wine. *The Use of a Calibration Sample in a Responsive Survey Design. American Association for Public Opinion Research.* FL: Hollywood, May 15, 2015.)

could either increase response rates or have little to no effect. However, an empirical test would be needed to evaluate the impact on cost, balancing the additional incentive cost with any reduction in interviewer time spent pursuing the nonrespondents.

There are also limitations to the use of trial and error. One-time surveys are limited in what can be tried, evaluated, and used to further optimize the design. Especially when prior data are not used, the decision of what alternatives to test can be very subjective—and there may be numerous design alternatives to be tried. Continuing the example with the NSFG nonresponse follow-up incentives, there are multiple incentive amounts that can be tested and the point at which the marginal returns begin to decline is unknown. The form of the incentive may interact with the incentive effect—some sample members may prefer a check, others cash, and yet others may prefer an electronic form such as gift card or code. Using past data to eliminate some designs and further inform the design being tested is in most cases necessary.

7.5 Summary

A key notion in ASD is that any ASD should be treated as suboptimal, and can be implemented with a plan to further improve the design—particularly in surveys that have replication (e.g., repeated cross-sectional surveys). We introduced the main approaches to optimization, and their main strengths and limitations

- *Mathematical and statistical optimization*. This approach can be used to optimize the design for multiple objectives. To do so, however, often strong assumptions need to be made. A challenge is posed by the need to estimate a large number of parameters, both computationally, as well as having a negative impact on the precision of these parameter estimates.
- *Simulation on existing data*. Particularly repeated cross-sectional surveys and surveys with a panel design have the benefit of simulating changes on prior data. This allows for evaluation of the effect of an ASD on other outcomes and testing multiple if–then scenarios. Nonetheless, many types of interventions cannot be simulated on past data without first embedding additional experiments.
- *Trial and error*. This approach should not be confused with trying a design without the ability to evaluate it. It simply means that there are still design choices for which the optimum solution is unknown, and experimental testing is needed. The design choices are usually informed by mathematical optimization, simulation, or past research, that is, not simply tests of randomly selected design alternatives. Trial and error can be used to improve ASDs across iterations, and it can be used in a separate evaluation. A key limitation of this approach is the number of alternatives that can be empirically evaluated. This is also why one may consider these three approaches in sequence, in the same order as presented here.

It is important to treat these approaches as complementary rather than alternatives. Each has unique strengths as well as limitations. For example, a complex optimization for multiple objectives can be achieved through mathematical optimization, the results can be validated through simulation on existing data that may identify any model misspecification, and finally, a single design alternative can be fielded and tested in data collection.

8

Robustness of Adaptive Survey Designs

8.1 Introduction

So far, we mostly ignored any inaccuracy in estimated survey design parameters such as contact propensities, participation propensities, costs per sample unit, or response quality propensities. Since these parameters form the main input to optimization of ASD, see Chapters 7 and 11, inaccuracy may lead to suboptimal performance or may even lead to ineffective designs. In this chapter, we discuss methodology to assess and to evaluate the impact of inaccurate parameters on the performance of designs.

Inaccuracy is the result of bias and/or sampling and estimation variance in estimated survey design parameters. Error may be introduced when parameters have been estimated using another survey, when parameters have been estimated on a nonrandom subset of the population, and/or when parameters have changed (gradually) in time. Error may also result from misspecified models for the parameters, that is, when some interactions between auxiliary variables are omitted or when an incorrect link function is used. We will not consider model misspecification as a source of bias here, but restrict the discussion to bias originating from the data that were used to fit the models. We refer to Chapter 5 for a discussion on modeling. It is important to note that in our context, bias is not introduced because individual probabilities or costs are replaced by stratum propensities or costs, but rather when stratum estimates are biased. In the optimization of ASD, we tailor and adapt to average nonresponse and measurement behavior in relevant strata based on auxiliary data. When two individual sample units have the same values on all auxiliary variables, then they are treated the same; for one individual unit the strategy may be more or may be less effective but it is the average effect that counts.

It is important to make a distinction between bias in ASD survey design parameters and bias in weighted estimates of survey outcome variables. Bias in survey design parameters used to assign a particular strategy does not necessarily result in bias in (weighted) survey estimates, as those are often used to assign a more effective strategy; the bias in the weighted survey estimates may not be affected, but the efficiency of the ASD may be reduced.

Sampling and estimation variance follows from the estimation of regression coefficients in models for survey design parameters from finite samples and from finite populations. Apart from censuses, survey design parameters are estimated on a sample of the population. Survey design parameter estimators are then subject to sampling variance, which is usually inversely proportional to the size of the sample. Estimation variance assumes an infinite or superpopulation model viewpoint and occurs even for censuses or surveys among the full population. Although we may fully observe the response behavior of a population at a certain time point and, consequently, have no sampling variance, the population may change, leaving us with uncertainty about the outcomes in a future survey. One may, for example, estimate the response propensity of a fully observed, but small, population stratum such as all students of a particular school or faculty. The fraction of students that responds is exactly known and has no sampling variance, but there still is a considerable uncertainty about the fraction of responding students in subsequent years when classes have changed. Estimation variance is usually inversely proportional to the population size. In practice, the two types of variance are often considered simultaneously and we will do the same here. The combined variance depends on the complexity of models for survey design parameters, that is, on the number of variables and their classifications, and the explanatory power of the auxiliary variables, that is, the homogeneity of strata formed by the auxiliary variables. For ASDs, there is a trade-off between precision of survey design parameter estimates and the number of the population strata. This trade-off depends to some extent on the strength of relations between the parameters and auxiliary variables. Hence, there is a close connection between stratification, see Chapter 3, and robustness of designs.

We have not defined yet, however, what we mean by robust ASDs. This could mean many things. It could mean little variation in realized costs, and/or quality, or a small probability that realized costs exceed a certain threshold or that realized quality remains lower than a specified threshold. Another possible definition is little systematic deviation to expected costs and/or quality. Finally, from the viewpoint of logistics and implementation, it could mean a stable allocation of data collection strategies to population strata, that is, few changes in the structure of the adaptation. These viewpoints represent different objectives that are at the core of ASD and that are deeply intertwined with the context and type of survey. See Chapter 2 for a more elaborate discussion. In some survey settings, it may simply not be acceptable that survey budget is overrun or that certain quality requirements, for example, the number of respondents or the response rate, are not met. Such settings demand little if no variation in costs and quality. They are more likely for occasional or one-time only surveys. ASD then needs to be dynamic and needs close monitoring of data collection. For some survey agencies and organizations it may be unacceptable that allocation schemes change constantly and require great flexibility in implementing data collection and assigning resources. In this chapter, we focus on systematic deviation from expected costs and/or quality, that is, a longer run friction between desirable and obtained quality and costs, and we consider stability in the allocation structure of ASD.

Obviously, a sensitivity to inaccurate design parameters, that is, a lack of robustness, is a feature that applies to all surveys, be they adaptive or nonadaptive. As explained, adaptation of data collection design may even be motivated by the need for stable quality and/or costs. It may seem counterintuitive that in general, parameter estimates in ASDs can be more sensitive to inaccuracy than nonadaptive designs as they employ detailed survey design parameters estimated for different population strata rather than for the whole population, but it is only under the expectation that the benefit of having multiple design parameters would be more beneficial for the properties of the survey estimates compared to no tailoring.

The robustness of ASDs strongly depends on the estimation strategy for survey design parameters. With unlimited resources, survey design parameters could be trained or learned and be constantly updated for a large range of possible design features, for example, multiple modes, contact protocols, or incentive amounts. Without logistical and implementation constraints, ASDs could then constantly, and organically, select the design features that lead to the most accurate statistics. Clearly, resources are limited, in fact they may be driving the search for efficient survey designs, and implementation requires some stability and predictability in order to avoid errors when executing the survey. Furthermore, a constantly changing survey design creates a perception of incomparability of survey statistics. Although this perception may be wrong, the population and survey climate in which the survey is conducted are not frozen and fixed, it usually forces survey redesigns to occur at a frequency of once in a number of years. As a consequence, the estimation strategy for stratum survey design parameters must be as efficient as possible and should lead to design decisions within a few months or quarters. Such a setting demands for conservative choices in ASDs, that is, accounting for the inaccuracy of survey design parameters. Such a setting also lends itself very well to a Bayesian framework in which data collection expert knowledge and historical survey data of the same, or comparable, surveys are included through prior distributions for survey design parameters. Efficient estimation strategies are, however, still in their infancy and a topic for future research and development.

The outline of the chapter is as follows. In Section 8.2, we introduce some metrics to assess the robustness of ASDs. In Section 8.3, we discuss and demonstrate sensitivity analyses employing these metrics. We briefly discuss a Bayesian framework for ASD in Section 8.4 that is currently being explored within an international network (Bayesian Adaptive survey DEsign Network or BADEN, see www.badennetwork.com). We end with a number of conclusions and recommendations in Section 8.5.

8.2 Metrics to Assess the Robustness of ASDs

We distinguish three viewpoints to robustness and link metrics to each of the viewpoints. The first and second viewpoints are relatively straightforward. The first viewpoint is that of variability in realized costs and quality. The obvious metric is the variance of the cost and quality criteria that are deemed important and drive the optimization of ASD, for example, the variance of the realized costs, the variance of the interviewer hours or calls, the variance of the response rate or numbers of respondents in relevant strata. The second viewpoint is that of systematic deviation in realized costs and quality. The obvious metric is then the bias of the same cost and quality criteria. Potentially, the two could be combined into a single metric such as the mean square error (MSE), but in practice it is useful to assess and evaluate them separately as they have different consequences and may require different countermeasures.

The third viewpoint, that of a stable allocation structure of strategies and population strata, is less straightforward and requires more elaboration and new metrics. Burger, Perryck, and Schouten (in press) proposed such metrics. We discuss them in the remainder of this section.

Before we introduce the design structure metrics, we introduce some notation. Suppose the population is stratified into G groups or strata, labeled as $g = 1,2, \ldots ,G$, based on auxiliary data that is linked from the sampling frame, administrative data, or paradata. Examples would be age or income groups or sample units with different interviewer observations. We consider a survey design d with one data collection phase with A possible actions or strategies, that is, $a \in \mathrm{A} = \{1,2, \ldots ,A\}$, for example, different interviewer modes or different contact strategies. Let $p_{d,g}(\varnothing)$ be the probability that a population unit in group g is not allocated to any action, that is, is not being sampled, for survey design d. This probability will, generally, be large in most surveys and is the complement of the sample inclusion probability. Let $p_{d,g}(a)$ be the probability that a population unit in group g is allocated to action a in survey design d, given that it is sampled. These strategy allocation probabilities sum up to one and describe the structure of the adaptation in the data collection phase. The population size for group g is denoted as N_g, the sample group size for design d is $n_g^{(d)} = N_g(1-p_{d,g}(\varnothing))$, and the total sample size for design d is $n^{(d)} = \sum_{g=1}^{G} n_g^{(d)}$. The relative sizes of group g in the population and sample are, respectively, $W_g = (N_g/N)$ and $w_g^{(d)} = (n_g^{(d)}/n^{(d)})$ for design d.

In the following, it is assumed that the stratification and possible actions are fixed before the data collection phase starts so that any design can be characterized by its strategy allocation probabilities.

We introduce three distance measures, labeled d_1 to d_3, to compare two designs, say design 1 and design 2. The three measures correspond to the three steps in the allocation of strategies to population units: determine the sample size, potentially vary sample inclusion probabilities over strata, and choose strategies per stratum. The measures take values in the interval [0,1], where a value 0 means that the designs are the same in a step and a value 1 means that the designs are optimally different.

The first measure is the absolute relative difference in sample size between design 1 and design 2

$$d_1(1,2) = \frac{|n^{(1)} - n^{(2)}|}{n^{(1)} + n^{(2)}}. \tag{8.1}$$

Measure d_1 is large whenever the two designs have very different sample sizes. The sum of the sample sizes in the denominator implies that the measure takes values smaller than 1.

The second measure is the (Euclidean) distance in group inclusion probabilities between design 1 and design 2.

$$d_2(1,2) = \frac{N}{n^{(1)} + n^{(2)}} \sqrt{\Sigma_g W_g (p_{2,g}(\emptyset) - p_{1,g}(\emptyset))^2}. \tag{8.2}$$

For simplicity, we assume that N and W_g are the same in both designs. d_2 is large whenever the two designs have very different probabilities to include population units from the different groups in the sample. In other words, after the sample is drawn the distribution of the groups differs between the two designs, for example, in the one design younger persons may be oversampled and in the other design older persons may be oversampled. d_2 is not affected by a difference in sample size, which is measured by d_1. So if all inclusion probabilities are the same, then it takes the value 0, regardless of sample sizes.

The third measure is the (Euclidean) distance in conditional allocation probabilities between design 1 and design 2

$$d_3(1,2) = \sqrt{\frac{1}{2G} \sum_{g=1}^{G} \sum_{a=1}^{A} (p_{1,g}(a) - p_{2,g}(a))^2}. \tag{8.3}$$

The scaling factor 1/2 in Equation 8.3 is again added to bound d_3 by 1. d_3 is large whenever groups get allocated very different strategies from one design to the other. For example, when younger persons are approached by Web and older persons by telephone in the one design and younger persons are approached by face-to-face interviewers and older persons by mail in the other design, then the designs are optimally different and d_3 equals one. When the survey mode is only different for older persons, then d_3 will be in between zero and one.

Together, the three measures tell us to what extent two (adaptive) survey designs are different. While two different designs may yield a similar performance in terms of quality and costs, they may pose very different demands in the implementation. From a logistical and operational viewpoint, we would like a design to be stable in time, that is, to have small d_1, d_2, and d_3 over time.

8.3 Sensitivity Analyses

In this section, we discuss three strategies to evaluate the sensitivity of designs to inaccuracy in survey design parameter estimators. Next, we apply the three strategies to an example linked to the Dutch LFS.

8.3.1 Strategies to Evaluate Robustness of Designs

We see four different, complementary options to evaluate robustness of survey designs; two based on bias in estimated survey design parameters and two based on imprecision of estimated survey design parameters. Survey design parameters are group response propensities, potentially decomposed to group contact and group participation propensities, and costs per sample unit in each group. These parameters are estimated from survey data from previous waves or previous data collection phases and are subject to bias and sampling variation. In a Bayesian setting, the survey design parameters may also be assigned prior distributions based on expert knowledge or estimated survey design parameters in other, similar surveys.

The four strategies are

1. Bias in survey design parameters
    a. Compare the performance in terms of quality and costs by adding bias to the survey design parameters
    b. Compare the structure in terms of measures d_1, d_2, and d_3 by optimizing (adaptive) survey design based on biased survey design parameters
2. Imprecision in survey design parameters
    a. Evaluate the variation in performance in terms of quality and costs by simulating values of survey design parameters that mimic sampling variation
    b. Evaluate the variation in structure in terms of measures d_1, d_2, and d_3 by optimizing (adaptive) survey design for different simulated values of survey design parameters that mimic sampling variation

A natural option to evaluate the impact of bias in survey design parameters is to compare the optimal design in one wave to that of a subsequent wave. To compare the performance, the (adaptive) survey designs are optimized in both waves. The optimal design of the first wave is applied in the second wave and response rates, costs, and other performance criteria are compared to that of the optimal design in the second wave. The difference in structure is assessed by computing the d_1, d_2, and d_3 for the optimal designs of the two waves.

Alternatively, when there is only one wave of the survey, one may add a bias to survey design parameters based on anticipated trends in survey design parameters, for example, gradually declining response propensities or rising costs.

A natural option to evaluate the impact of sampling variation is to resample the survey data and re-estimate the survey design parameters on each resample. Resampling may be done using bootstrap, in which the sampling design is simulated. For each resample, the variation in response rates, costs, and other criteria is assessed for the optimal (adaptive) survey design. For each resample, the (adaptive) survey design can also optimized again, and the structure of the optimal designs may be compared through the d_1, d_2, and d_3.

Alternatively, when resampling is complicated or impossible due to the absence of complete survey data, one may formulate a simulation model for the survey design parameters and repeatedly draw values for which performance and structure is assessed.

The impact of bias and imprecision may form the input to design decisions themselves. One may choose (adaptive) survey designs that are less sensitive to anticipated or known changes and variation in survey design parameters.

8.3.2 The Dutch LFS: An Example

We return to the Dutch LFS to demonstrate how robustness can be analyzed. We simplified the example somewhat, so that it becomes possible to evaluate all candidate strategies directly and no numerical optimization is needed. The LFS has a sequential mixed-mode design, Web followed by either telephone or face-to-face. Web nonrespondents are called when a phone number is available and are visited at home otherwise. In the example, we take the two interviewer modes together and decide only whether we follow-up Web nonresponse with yes or no, that is, there are two phases with two possible actions in phase 2: an interviewer follow-up or no action. We distinguish three population strata based on age: persons 15–24 years, 25–44 years, and 45–65 years. Furthermore, a paradata variable becomes available during the Web first phase: a binary indicator for a break-off (yes or no). We use only 0–1 allocation probabilities per stratum to the interviewer follow-up. Consequently, there are $2^6 = 64$ possible designs (six strata and in each stratum a follow-up is applied or not). Tables 8.1 and 8.2 give the response propensities and costs per sample unit for each of the six strata in two consecutive years, say 2016 and 2017. The relative sample sizes of the six strata are shown in Tables 8.1 and 8.2.

TABLE 8.1

Stratum Response Propensities for Two Phases in the LFS in Two Consecutive Years t and $t+1$

			Year 2016		Year 2017	
Age	Break-off	Relative size (%)	Phase 1	Phase 2	Phase 1	Phase 2
15–24	No	25	0.20 (0.02)	0.40 (0.04)	0.18	0.40
	Yes	8	0	0.70 (0.04)	0	0.70
25–44	No	28	0.30 (0.02)	0.60 (0.04)	0.25	0.50
	Yes	5	0	0.80 (0.04)	0	0.75
45–65	No	25	0.40 (0.03)	0.50 (0.04)	0.40	0.40
	Yes	9	0	0.80 (0.04)	0	0.75

Note: For the first year, standard deviations are given for the response propensities. Third column gives relative sizes in the population.

We assume that 2016 data are used to optimize the LFS ASD. Therefore, for 2016, response propensities and costs per sample unit, standard deviations are given in Tables 8.1 and 8.2. The standard deviations represent the uncertainty in the "true" values and are used to draw different (independent) samples and mimic sampling variation. The response propensities are drawn from a Beta(α,β) distribution. The two parameters are chosen such that the expectation and variance match the values in Table 8.1. The costs per sample unit are drawn from a normal distribution $N(\mu,\sigma^2)$, where μ and σ are given in Table 8.2. The year 2017 response propensities and costs per sample unit are used to mimic time change and bias.

Now we have set the simulation framework, we need to specify cost and quality criteria. In the example, we maximize the response rate subject to constraints on the total costs, the number of respondents in the three age strata, and the variation in the response propensities measured through the R-indicator (see Chapter 9). The budget for the LFS is 40,000 EU, the

TABLE 8.2

Stratum Costs Per Sample Unit for Two Phases in the LFS in Two Consecutive Years t and $t+1$

			Year 2016		Year 2017	
Age	Break-off	Relative size (%)	Phase 1	Phase 2	Phase 1	Phase 2
15–24	No	25	4 (0.1)	25 (2)	4	30
	Yes	8	5 (0.1)	30 (2)	5	32
25–44	No	28	4 (0.1)	20 (1)	4	22
	Yes	5	5 (0.1)	25 (2)	5	26
45–65	No	25	3 (0.1)	15 (1)	4	16
	Yes	9	5 (0.1)	20 (2)	5	21

Note: For the first year, standard deviations are given for the costs. Third column gives relative sizes in the population.

minimum expected number of respondents in each stratum is 500, and the expected *R*-indicator for age should be at least 0.80.

The LFS budget is too small to perform an interviewer follow-up for all Web nonrespondents. The minimal numbers of respondents can easily be met by conducting only the Web data collection phase, since this phase is relatively cheap. However, the *R*-indicator requirement forces the design to balance response by allocating some of the Web nonrespondents to the interviewer second phase. Bias in response propensities and/or costs, for example, overestimating the Web response propensities, may lead to a suboptimal design or even to a design that is not meeting the requirements. Imprecision in response propensities and/or costs leads to variation in performance of the design and may imply that for some realizations costs are too high or respondent numbers are below the thresholds. The latter may, of course, occur equally well in uniform designs, but in ASD we consider more detailed and, hence, more noisy survey design parameters.

We analyze robustness by considering the two bias and variance viewpoints of Section 8.3.1.

First, we evaluate the 64 possible strategies using the expected response propensities and costs for 2016. Seven of the 64 strategies satisfy the *R*-indicator and cost constraints, when the sample size is chosen such that the 500 respondents per stratum are obtained (in expectation). The strategy with the highest response rate out of these seven is the strategy that does an interviewer follow-up for three strata, namely 15–24 years and break-off, 25–44 years and break-off, and 45–65 years and no-break-off. The expected response rate of this design is 40.5%, the expected *R*-indicator is 0.836 and the expected required budget is 38.320. Figure 8.1 gives a boxplot for the

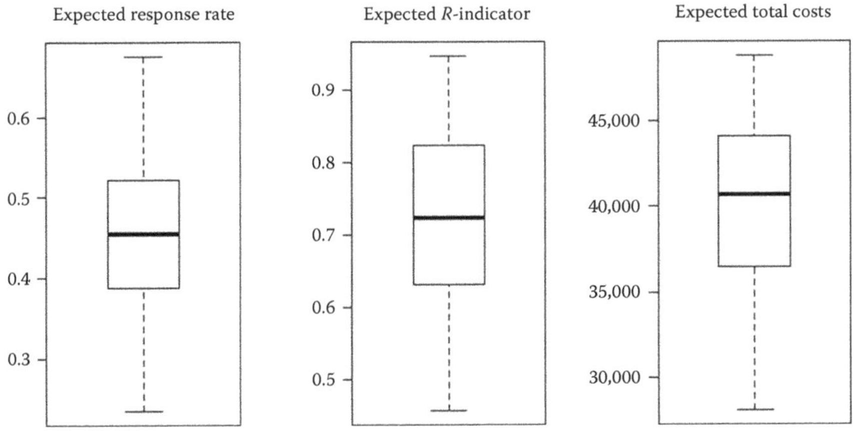

FIGURE 8.1
Boxplots for expected response rates, *R*-indicators, and total costs for the 64 possible strategies in 2016.

expected response rates, R-indicator and costs of all 64 strategies. The median R-indicator is 0.72 and the median total cost is 41,000 EU. Under the 64 possible strategies some almost achieve perfect balance, an R-indicator close to 1 or costs that are 10,000 EU below budget. However, these do not satisfy other criteria.

We analyze the impact of imprecision on performance by sampling 1,000 sets of response propensities and costs per sample units for 2016 using the beta and normal distributions. Next, we compute corresponding expected response rates, R-indicators and costs for the optimal strategy. Figure 8.2 gives the histograms for the response rates, R-indicators, and total costs. The variation in response rates is relatively modest, but for the R-indicators and costs is relatively large. About 5% of the simulated R-indicators fell below the 0.8 constraint, whereas 16% of the simulated costs were larger than the budget of 40,000 EU. The conclusion is that the quality and costs of the optimized design is sensitive to the imprecision in the response and costs parameters.

We analyze the impact of imprecision on structure, by again sampling 1,000 sets of response propensities and costs per sample unit for the two phases, and by optimizing the design for each set. We compare the structure of the designs with the optimized design based on the original set of response propensities and costs, that is, the design with a follow-up for persons 15–44 years with break-off and 45–65 years without break-off. Figure 8.3 shows histograms of the three metrics d_1 (difference in sample size), d_2 (difference in group sampling probabilities), and d_3 (difference in group strategy allocation probabilities) for the 1,000 simulated sets of parameters. The sample sizes vary relatively little, see the histogram for d_1; 65% deviate less than 3%

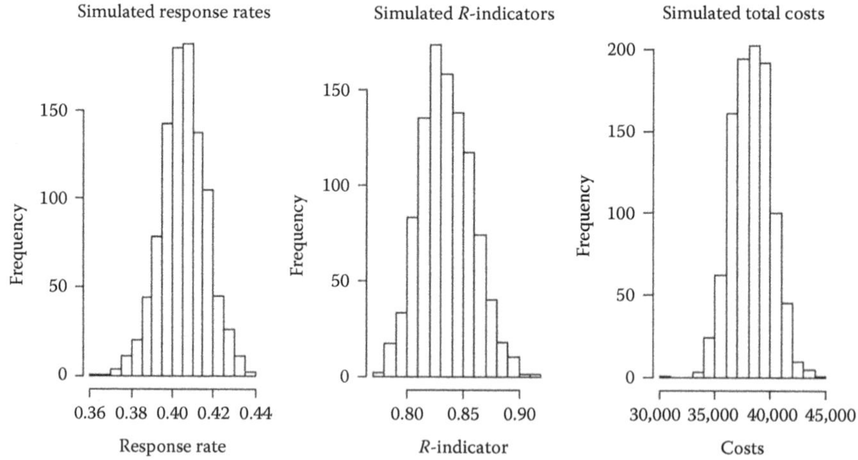

FIGURE 8.2
Histograms for the response rate, R-indicator, and total costs for the 1,000 samples from the optimal design in 2016.

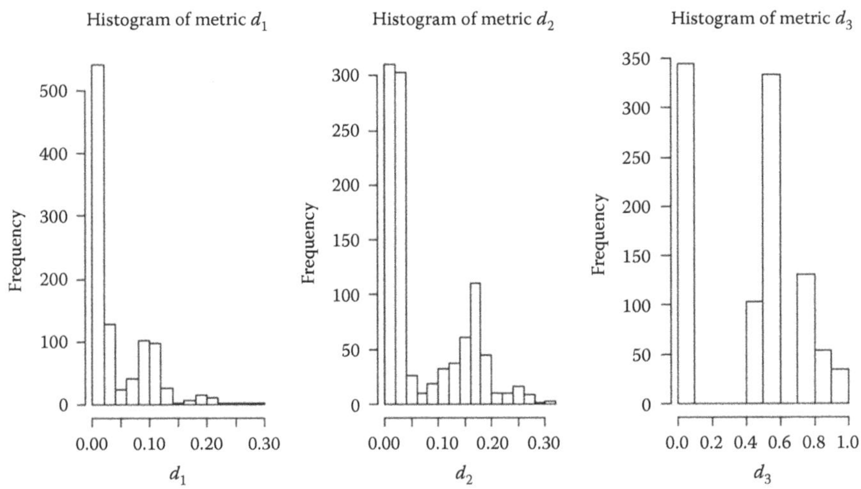

FIGURE 8.3
Histograms for the three metrics d_1, d_2, and d_3 for 1,000 simulated sets of parameters in 2016.

from the original design and 85% deviate less than 10%. The group sampling probabilities, see the histogram for d_2, vary somewhat more, but still more than 50% deviate less than 3%. However, the d_3 can attain rather large values. Out of the simulated sets of parameters 34% led to exactly the same strategy allocation, 43% changed only one follow-up stratum, 18% changed two follow-up strata, and 4% changed three follow-up strata. Hence, we can conclude that sampling variation leads to a relatively stable sampling design but greatly varying strategy allocations.

Finally, we consider the impact of bias by comparing the optimal 2016 and 2017 designs. We compare the response rate and metrics d_1, d_2, and d_3 for the optimal 2017 design to the 2016 design. The optimal design for 2017 allocates a follow-up to two strata, namely to 15–24 years with a break-off and to 25–44 years with a break-off. Hence, with respect to 2016, the follow-up for persons 45–65 years without break-off is no longer allocated to the interviewer mode. This can be explained by the lower online response rates which imply that a larger budget is needed; only four out of the 64 strategies satisfy all constraints. The metrics for the distance between the optimal 2016 and 2017 designs equal to $d_1 = 0.02$, $d_2 = 0.15$, and $d_3 = 0.41$. So the sample sizes are quite similar, but the strata allocation probabilities have changed because stratum 45–65 years without break-off no longer receives a follow-up, and, hence, has a lower expected response propensity. The budget required to do the optimal 2016 design in 2017 would need to be 44,053 EU, that is, should increase by around 10% due to the lower online response rates. From these simulations, we can conclude that the optimal design is quite sensitive to bias in response and costs parameters. Figure 8.4 contains the boxplots for the expected response rates, *R*-indicators, and costs for the 64 possible strategies

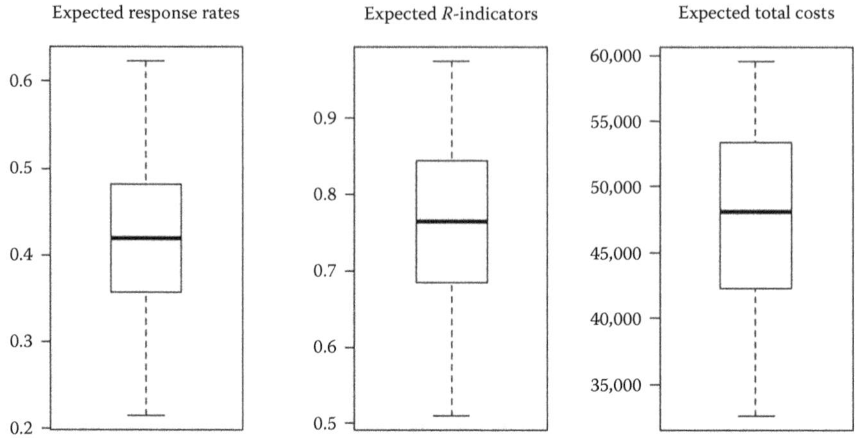

FIGURE 8.4
Boxplots for expected response rates, *R*-indicators, and total costs for the 64 possible strategies in 2017.

based on 2017 parameters; the median response rate and *R*-indicator did not change much but the median budget went up from 41,000 to 48,000 EU.

The analyses provide insight into the impact of inaccuracy on performance and structure. The results may make for a stepping stone to design final decisions. When it is deemed imperative that an overrun of budget does not occur, then a constraint on the probability of budget overrun can be used instead of the budget constraint itself. Alternatively, when the minima numbers of respondents are crucial, then the probability of a shortage of respondents can be limited. These constraints will, generally, lead to more conservative designs.

8.4 Bayesian Adaptive Survey Design Network

In 2015, a 3-year international network, Bayesian adaptive survey design network (BADEN) started to explore the potential of Bayesian analyses in ASD. The rationale behind this initiative is that historic survey data of the same or similar surveys, together with expert knowledge of data collection staff, may facilitate the optimization given inaccurate survey design parameters and increase efficiency in learning and updating survey design parameters. Such historic data and expert knowledge may be incorporated in a natural way in a Bayesian analysis. See Bruin et al. (2016) for details. In this final section, we present some early ideas and considerations.

In a Bayesian analysis, survey design parameters are not viewed as fixed quantities but are random variables that have a prior distribution before data collection starts and a posterior distribution when data collection is completed. The posterior distribution can be seen as updated knowledge about the potential values of the survey design parameters and is proportional to the product of the prior distribution and the likelihood of the observed data.

Let us again consider the two phase LFS example and suppose that response propensities and costs per sample unit are modeled using main effect regression based on the person's age in three classes and the break-off indicator. The model may then be written as

$$\tilde{R}_{t,i} = \alpha_{t,R}a_i + \beta_{t,R}b_i + \varepsilon_{t,R,i}, \tag{8.4}$$

$$R_{t,i} = \begin{cases} 1 & \text{if} \quad \tilde{R}_{t,i} \geq 0 \\ 0 & \text{if} \quad \tilde{R}_{t,i} < 0 \end{cases}, \tag{8.5}$$

$$c_{t,i} = \alpha_{t,C}a_i + \beta_{t,C}b_i + \varepsilon_{t,C,i}, \tag{8.6}$$

where a_i is the age of person i, b_i is the break-off indicator for person i, $\tilde{R}_{t,i}$ is a latent response tendency of person i in phase t, $c_{t,i}$ is the costs per sample unit for person i in phase t, and $\varepsilon_{t,R,i}$ and $\varepsilon_{t,C,i}$ are independent normally distributed error terms with zero expectation and variances equal to 1 and σ_t^2, respectively. The nonresponse model amounts to a probit regression where the binary response indicator $R_{t,i}$ equals one whenever the response tendency exceeds 0. The response propensity of person i in phase t can be derived as $\rho_{t,i} = \Phi(-\alpha_{t,R}a_i - \beta_{t,R}b_i)$, where Φ is the probability distribution function of the standard normal distribution.

In a Bayesian analysis, the regression coefficients $\alpha_{t,R}$, $\beta_{t,R}$, $\alpha_{t,C}$, $\beta_{t,C}$, and σ_t^2 are assigned prior distributions that are derived from historic data and/or expert knowledge. The regression coefficients over different phases may be jointly modeled in a further step when they are deemed dependent, which introduces an additional modeling level and leads to a so-called hierarchical model. We do not elaborate such an extension here. Natural choices for the prior distributions for the regression slope parameters $\alpha_{t,R}$, $\beta_{t,R}$, $\alpha_{t,C}$, and $\beta_{t,C}$ are normal distributions. For the regression dispersion parameters σ_t^2 choices may be inverse gamma or Weibull distributions.

When assigning prior distributions to the regression coefficients, indirectly, aggregate quality and cost functions such as the (subgroup) response rates, coefficients of variation of the response propensities and total costs, also get prior distributions. These may, however, be complex and exotic in form, but essentially are the distributions of interest in the analysis.

There is extensive literature on how to derive posterior distributions from the prior distributions and probability likelihood linked to the observed data, for example, see Gelman et al. (2014) for an overview of the literature and methods. In general, in the sketched framework for survey data collection analysis, the posterior distributions cannot be written in a closed form—the distributions contain components, usually integrals that cannot be evaluated analytically. Consequently, posterior distributions need to be derived by approximation. There is a wide range of methods to do this, but the most popular is through so-called Markov Chain Monte Carlo (MCMC), and more specifically through Gibbs samplers. These methods produce dependent draws from the posterior distributions of interest. They require one or more starting points and, therefore, have a so-called burn-in period. Execution of the methods can be time-consuming but standard routines are available in packages such as R and SAS. Expectations, variances, quantiles, and other properties of the posterior distributions can be estimated from the MCMC draws by the empirical distributions.

The posterior distributions form the starting point for design decisions. They may, for example, state that the coefficient of variation with respect to age and break-off is expected to be 0.20 with a standard deviation of 0.05 when all sample persons receive a follow-up with an interviewer. In order to make such choices adaptive, the models for response and costs should contain the auxiliary variables of interest, in the example age and break-off. Posterior distribution for the coefficient of variation and other aggregate functions may be derived from the same MCMC draws for designs that discard interviewer follow-up for some of the age and break-off strata.

Apart from the inclusion of historic data and expert knowledge, the posterior distributions offer the benefit of expressing uncertainty in the design decisions. Depending on the amount of information in the prior distributions and the observed data, the posterior distributions have a smaller or larger variance. They may, for instance, state that the probability that budget is exceeded under a certain design is 15%. They also provide probabilities that an ASD, for example, performing a follow-up only for younger persons and persons that broke off, is performing better than a nonadaptive design. The posterior distributions may be used as prior distributions to new waves or cycles of the same or similar surveys.

The most important element of the Bayesian analysis is the elicitation of prior distributions from the same or similar surveys. Such elicitation is not entirely straightforward for two reasons. First, data collection experts may have a relatively good sense about response propensities and costs for certain population subgroups, but not about coefficients in regression models. An intermediary step is needed to translate such knowledge, for example, by fitting regression models to simulation data based on expert experience. Second, in time, survey design parameters change but the Bayesian analysis does not account for that without any modification. When the observed data are large in size, then the posterior distribution obtains a very small variance and is concentrated around the mean. Feeding such a posterior distribution to a new wave of data collection is meaningless as it can no longer move or change. A solution is to moderate prior knowledge depending on the time it was observed. Hence, prior elicitation needs to be done with care and requires some experience by itself.

Finally, the Bayesian analysis may, and perhaps should, be extended with a model for the main survey variables, especially when a survey has a relatively small set of topics and key statistics. In addition to (8.3) through (8.5), there are regression models for a survey variable, say a continuous variable y_i. Since a change of mode may lead to another selection of respondents and to a different measurement, the data collection phase should be included in the model

$$y_{t,i} = (\alpha_Y + (t-1)\Delta_Y)a_i + \beta_Y b_i + \varepsilon_{t,Y,i}\,, \tag{8.7}$$

where Δ_Y is the vector of mode effects that should be assigned a prior distribution as well.

8.5 Summary

We make a number of observations related to robustness of ASD:

- ASDs are more sensitive to inaccuracy in survey design parameters when stratification becomes more detailed, that is, it is imperative to be modest in the number of strata that is distinguished.
- While robustness to inaccurate survey design parameter estimates is a feature of any survey design, ASDs are more sensitive than nonadaptive survey designs, in general, as they detail estimates to population strata.
- Inaccuracy in survey design parameters results from sampling and estimation variance, bias due to temporal change, for example, declining response propensities, and bias due to the use of experimental data or data from similar, but different surveys.
- There are three main viewpoints to robustness that require different metrics, analyses, and countermeasures: (1) no or little variation in realized costs and quality, (2) no or little systematic deviation in realized costs and quality, and (3) stable logistics and implementation through a constant strategy allocation structure.
- Empirical evidence seems to suggest that ASD is most sensitive to change/bias in the location of survey design parameters and less so to imprecision.
- A Bayesian analysis may prove very useful to express uncertainty and account for uncertainty in the optimization of ASD, but, to date, has received little attention. A crucial element is the elicitation of prior distributions from expert knowledge and historic data of the same or similar surveys.

Section IV

Advanced Features of Adaptive Survey Design

9

Indicators to Support Prioritization and Optimization

9.1 Introduction

ASD found its origin in attempts to reduce the impact of nonresponse bias on survey statistics. Since nonresponse bias is unknown for the survey outcome variables of interest, the survey literature has put forward various indicators that are employed as surrogates for nonresponse bias. ASD may consider such indicators as quality criteria in making decisions or optimizing the allocation of resources. In this chapter, we present an overview of the indicators that are used most often and we discuss their statistical properties. The indicators come from different survey contexts and from different modeling paradigms. As a consequence, they have rarely been discussed in a single document.

The indicators typically depend on auxiliary information, and without specifying what auxiliary variables are included and how, the indicator values are meaningless. Given that the values of the indicators are inseparable from the auxiliary variables that are included, their values mostly have a relative rather than an absolute meaning. That is, while they may give an indication of issues of performance relative to auxiliary variables, they do not provide an absolute estimate of the bias. In this chapter, we will not discuss the search for sensible auxiliary variables, which is the topic of Chapter 3. However, since ASD is essentially a form of nonresponse adjustment by design, just like traditional adjustment, useful auxiliary variables need to relate both to response behavior and to the survey variables of interest.

The indicators also typically need to be estimated using samples drawn from a population. In this chapter, we will only briefly mention the estimation of indicators and refer to Chapter 5 for details about modeling nonresponse. Since indicators are estimated from samples drawn from a larger population, they have a certain imprecision. This imprecision is often neglected, but for surveys with small samples this imprecision plays a dominant role; moderate to large samples are needed to allow for strong statements about the impact of nonresponse. Also the archetype indicator, the response rate, is usually presented without error margins.

Nonresponse bias is often used as a general feature or property of a survey, but it is unspecified until population parameters and corresponding estimators are chosen. To date, most literature has focused at the mean or total of survey variables in the population. We will do the same here, but there is a clear need for a more general view, given that many survey users are interested in associations between two or more variables rather than the levels of the individual variables.

In Chapters 10 and 11, we will include control of measurement error and sampling error as objectives in ASDs. Some of the design choices and interventions available to ASDs may have a big impact on answering behavior and data quality, so that measurement error needs to be accounted for. Furthermore, as mentioned, ASD is an adjustment for nonresponse by design; it, therefore, impacts precision of statistics and it needs to be combined with other nonresponse adjustment methods that are made afterward.

To precisely define and discuss the various indicators, a common, overarching notation is needed and the theoretical nonresponse bias of estimators for population means needs to be derived. Such an exercise is very statistical and may not be needed to use the indicators in practice. For this reason, we divide the chapter into two parts. In the first part, Sections 9.2 and 9.3, we loosely define the indicators and demonstrate their use. In the second part, Sections 9.4 and 9.5, we define the indicators in statistical notation and provide in-depth derivations of nonresponse bias and how the indicators appear in such expressions. The second part may be skipped. We will use a simple example to illustrate our definitions and to clarify meaning throughout the chapter.

EXAMPLE 9.1: LFS

The example in this chapter is a simplified version of the example that is used throughout the handbook in order to illustrate the use of indicators. An important statistic produced by LFS is the proportion unemployed in the labor force population, persons between 15 and 65 years of age. Households are recruited through an invitation letter with a login to a Web survey. Households not responding to the request receive a follow-up by a face-to-face interviewer. In the face-to-face follow-up, a standard effort and an extended effort are distinguished, where extended effort implies a longer fieldwork period and extra interviewer visits. Based on sampling frame data, the population is divided into two strata: between 15 and 25 years of age, and older than 25 years of age. During the Web first phase, paradata are collected about break-offs, that is, persons entering the Web questionnaire but quitting without submitting the questionnaire. At the first face-to-face visit, the interviewer makes an observation on the state of the dwelling, whether the dwelling shows some deterioration. The indicator for a break-off forms two strata that can be used in the face-to-face follow-up, and the interviewer

TABLE 9.1

Population Sizes, Response Rates and Unemployment Rates for the LFS Example

Age Category	Web Break-off	Dwelling Bad Shape	Stratum Size (%)	Response Web (%)	Response F2F Standard (%)	Response F2F Extended (%)	Unemployment Rate
>25	No	Yes	30	35	50	60	2
>25	No	No	20	20	40	50	5
>25	Yes	Yes	5	0	80	90	5
>25	Yes	No	5	0	75	85	8
15–25	No	Yes	20	25	35	50	12
15–25	No	No	15	15	20	35	25
15–25	Yes	Yes	1	0	75	80	14
15–25	Yes	No	4	0	70	75	30
All	All	All	100	22	44	56	10

observation gives an additional two strata that can be used to decide on an extended face-to-face strategy. Taken together, eight strata arise from age, break-off, and dwelling status. The relative size, the response rates, and the unemployment rate per stratum and for the total population are given in Table 9.1. The unemployment and break-off rates are set slightly higher than reality in order to make the impact on nonresponse bias more pronounced.

9.2 Overall Indicators

Nonresponse indicators can be divided into two main types, following the typology presented by Wagner (2012): indicators that include only observed auxiliary variables, and indicators that also include observed survey variables. Although the selection of auxiliary variables may be based on their association to the survey variables, the type 1 indicators typically attempt to provide a general view on the consequences of nonresponse, while type 2 indicators are specific to survey variables and are (usually) based upon multivariate relationships. We will discuss them separately, but we will present the relations between them in Sections 9.4 and 9.5. Schouten, Calinescu, and Luiten (2013) term the two types of indicators covariate-based and item-based, where item refers to the survey items in the questionnaire.

9.2.1 Type 1 Indicators (Based on Covariates Only)

The current literature on nonresponse indicators has focused attention mostly on the association with bias of estimators for means and totals. The traditional response rate, usually estimated by the design-weighted mean of the individual 0–1 response indicators, is an exception. The response rate is a component of the bias of many, if not all, estimators and population parameters. Schouten, Cobben, Bethlehem (2009) supplement the response rate by general indicators that measure the distance of response propensities to the response rate. Their motivation comes from the observation that variance of the response probabilities, $S^2(\rho)$, has a general impact on the bias of many estimators. In fact, when $S^2(\rho) = 0$, then nonresponse may be viewed as just a random second stage to the sampling design. Since the true response probabilities are unknown, they are replaced by response propensities defined for population strata based on auxiliary information X and transform the resulting variance, $S(\rho_X)$, to the 0–1 interval. The resulting indicator is termed the representativeness indicator, or simply R-indicator, and is defined as

$$R(X) = 1 - 2S(\rho_X), \tag{9.1}$$

which equals one when the response propensities are constant. The lower bound to the R-indicator is

$$1 - 2\sqrt{\bar{\rho}(1-\bar{\rho})}, \tag{9.2}$$

which equals 0 when $\bar{\rho} = 0.5$. Expression (9.2) shows that indeed the R-indicator is only a supplemental measure as the lower bound increases when the response rate decreases. For this reason, Schouten, Cobben, and Bethlehem (2009) propose to combine the response rate and the R-indicator through the nonresponse bias of design-weighted response means as (naive) estimators for population means. Furthermore, the reference to the auxiliary variable vector X in $R(X)$ is crucial. If this model is incorrectly specified (i.e., the "wrong" vector X is chosen), then the indicator may not be a good predictor of when bias is likely to happen and may lead to misleading conclusions.

How can the R-indicator be interpreted? Assuming that all of the variables in the vector X are categorical, then the R-indicator can be seen as a measure of the variability of subgroup response rates. The more these subgroup response rates vary, the lower the R-indicator goes. The concern is that if the response process is selecting a set of cases that are different from the sample with respect to X, then they may also differ with respect to the survey outcome variables. For example, if two key strata are defined by age, and the older age group responds at a much higher response rate than the younger age group, then it might be that the resulting estimates from the survey are biased, even after adjusting for the differences in response rate. A more formal description of the circumstances under which this is true can be found in Section 9.4.

Another, related indicator is the coefficient of variation of the subgroup response rates $CV(X)$. This indicator is similar to the R-indicator in that it is based upon the variation of subgroup response rates. This variation is standardized by dividing by the response rate.

$$CV(X) = \frac{S(\rho_X)}{\bar{\rho}}. \tag{9.3}$$

The logic of this indicator is that variation in subgroup response rates is an indicator of a selective response process that may lead to bias with respect to the survey outcome variables. Like the *R*-indicator, the choice of the subgroups plays a crucial role in determining the validity of the measure as an indicator of nonresponse bias. A more detailed description of this indicator can be found in Section 9.4.

Other indicators of this type have been developed. The imbalance, IMB(*X*), and distance, dist(*X*), indicators that are introduced by Särndal (2011), relate strongly to the *R*-indicator and coefficient of variation. We will demonstrate the link in Section 9.5 (Figure 9.1).

EXAMPLE: LFS

As a reminder, this survey proceeds through three phases of data collection. The first phase is a Web survey. The second phase is the standard face-to-face data collection. The third phase is an extended face-to-face data collection with additional call attempts. There are three auxiliary variables of interest: $X_{1,i}$ is the 0–1 indicator for being between 15 and 25 years of age of person *i*, $X_{2,i}$ is the 0–1 indicator for a break-off during Web of person *i*, and $X_{3,i}$ is the 0–1 indicator for a normal dwelling status for person *i*. In the monitoring of data collection, two questions are asked:

1. How does the representation of the age (X_1) and break-off strata (X_2) evolve from phases 1–3?
2. How does the representation of all strata change in the interviewer phases?

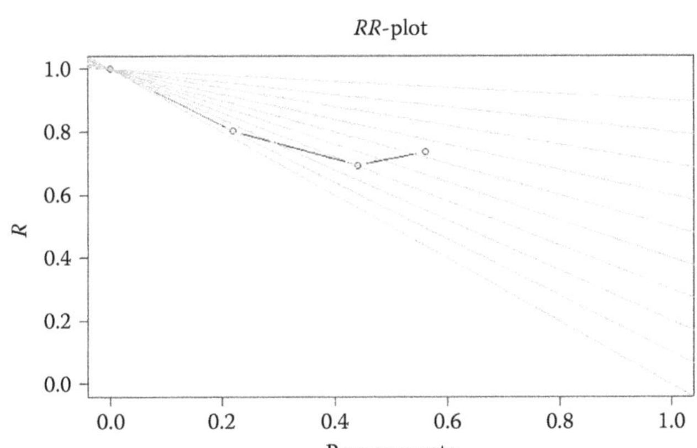

FIGURE 9.1
Plot of *R*-indicator against response rate after each phase for the model $X_1 \times X_2$. Diagonal lines represent constant coefficients of variation.

$R(X)$ and $CV(X)$ are computed for two different models at the end of each phase. Table 9.2 shows the values of the indicators. The results from the $X_1 \times X_2$ model are in the first three columns of reported values. This model excludes the variable indicating whether there was a break-off during the Web survey since it is only available after the Web phase. The last two columns show the resulting values for the indicators under the model $X_1 \times X_2 \times X_3$ after the second and third phase, conditionally on being a phase 1 nonrespondent. To answer the first question: The R-indicator shows that variation in response was lowest after the first phase (Web) and improves in phase 3 (extended) relative to phase 2 (standard). The CV shows a steady improvement (lowering) from phases 1–3, and suggests that representation is best at the end of data collection when the interest is in totals and means. Hence, one may conclude that the increase in response rates in the interviewer phases outweighs the increase in response propensity variance after the first phase; the risk of nonresponse bias has decreased. The answer to the second question: Both indicators show an improvement in overall representation in the extended interviewer phase. Figure 9.2 gives a plot for R against the response rate, a so-called RR-plot. Diagonal lines correspond to fixed CV values.

EXAMPLE: NSFG

The NSFG also tracks the R-indicator and the CV. The model for the R-indicator was selected at the beginning of production. Two different models are estimated. The first model is for response to the screening survey. The second model is for response to the main interview and is estimated only for cases that have been identified as eligible from the screening interview. The predictors for the two models differ somewhat since the model for the main interview can include variables obtained from the screening interview. The predictors for the model for the screening interview include the Census Region, the occupancy rate for the Census Block Group of the sampled housing unit estimated from ACS, the age eligibility rate of the population for the Census Tract of the sample housing unit estimated from the ACS, interviewer observations about whether the neighborhood is entirely residential or a residential/commercial mix, is there evidence of non-English speakers in the neighborhood, are there safety concerns in the neighborhood, and housing

TABLE 9.2

Various Type 1 Indicators for LFS Example

	$X_1 \times X_2$			$X_1 \times X_2 \times X_3$	
		F2F		F2F	
	Web	**Standard**	**Extended**	**Standard**	**Extended**
$R(X)$	0.802	0.694	0.737	0.527	0.593
$CV(X)$	0.454	0.345	0.236	0.822	0.468

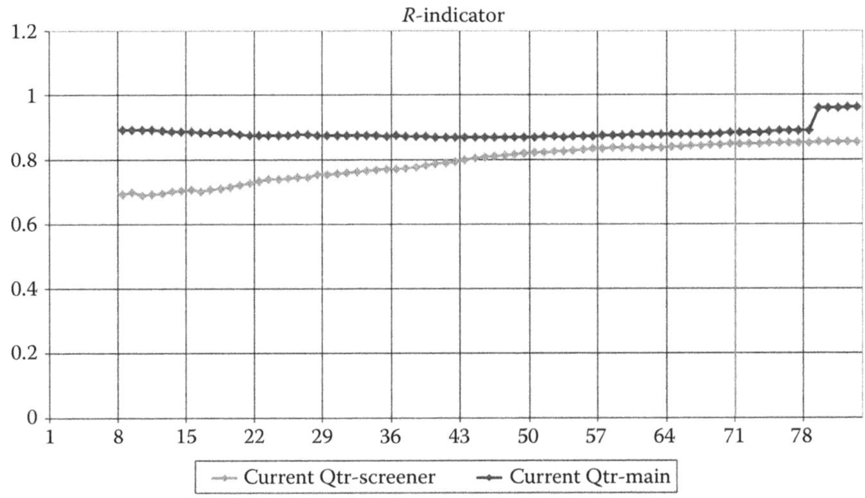

FIGURE 9.2
R-indicator for the NSFG screener and main models.

unit observations such as are there any impediments to accessing the housing unit, is the housing unit a single family home or a multi-unit structure, and a measure of urbanicity derived from information about the county of the sampled housing unit. The main interview model includes the variables from the screening interview model, and adds several variables derived from the screening interview, including the age, sex, and race of the sampled person, household size, and the interviewer observation about whether the selected person is in a sexually active relationship with a person of the opposite sex.

Figure 9.2 is an example from the NSFG dashboard that is used to track the *R*-indicator over the 12-week field period. The lower line is the *R*-indicator for the screener model. The upper line is the *R*-indicator for the main model. The vertical line at 70 days represents the boundary between phases 1 and 2.

The NSFG also monitors the *CV* of the response rates of key demographic subgroups. These subgroups define important analytical domains. Further, the demographic variables that define these subgroups are important predictors of many of the survey outcome variables, which are related to family formation, fertility, and sexual relationships. The groups are defined by the cross-classification of two age categories (15–19 and 20–44), two sex categories (female and male), and three race/ethnicity categories (non-Hispanic black, Hispanic, and other). The cross-classification of these three variables produces 12 subgroups.

Figure 9.3 shows the *CV* of these subgroup response rates by day. The *CV* for the current quarter is the line with a diamond at the point representing the *CV* for that day in the field period. Included in the figure are several comparison lines that are based on previous quarters, years, or cycles of the NSFG.

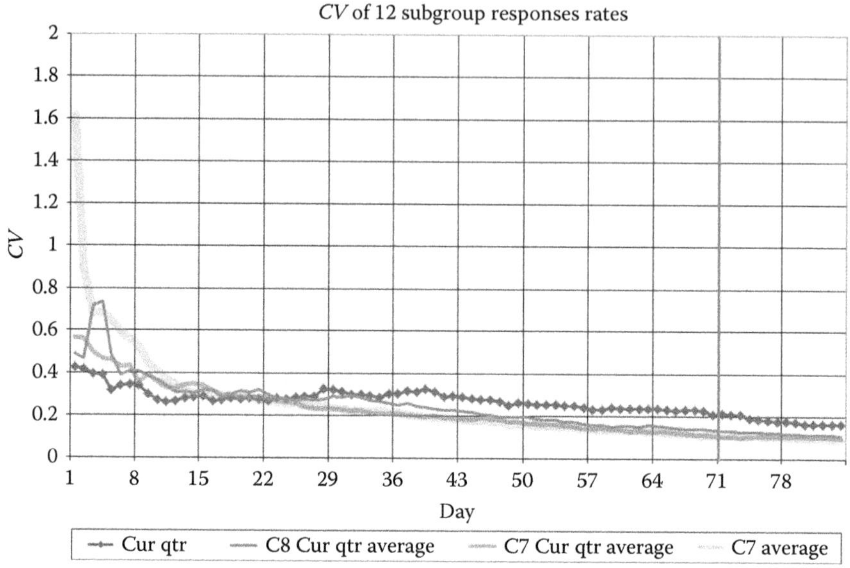

FIGURE 9.3
CV of NSFG subgroup response rates.

So far, we have ignored that response propensities are unknown. In practice, this does not hold and response propensities need to be estimated from survey response. Chapter 5 is devoted to models for nonresponse and we will not discuss estimation here in detail. However, a few comments are in order.

The estimation of indicators from estimated response propensities and from the design-weighted sample introduces inaccuracy. The sample-based estimators for the indicators are subject to sampling variation and they need to be accompanied by confidence intervals. For example, it is possible to construct confidence intervals around the *R*-indicator. The sampling error can be quite large. It is necessary to consider this variation when considering whether to base actions upon the *R*-indicator. Furthermore, a sample-size dependent bias is introduced due to the quadratic forms of the indicators. For moderate sample sizes, $n = 10{,}000$ and larger, this bias is, usually, negligible. We refer to Shlomo, Skinner, and Schouten (2012) and De Heij, Schouten, and Shlomo (2015) for details about the estimation of the indicators and their variance.

In order to compare indicator values it is imperative that the model used for estimating response propensities is kept fixed. That means that the same variables need to be included, with the same classifications if they are categorical, that the same main and interaction effects need to be included, and that the same link function, for example, linear, logistic, or probit, is applied. Hence, variables need not be selected based on automated test procedures run at each point in time, but instead need to be included at every time point. This insures comparability across time points. Further, since small samples may not allow detailed models to be estimated due to numerical and convergence problems, it is important that some parsimony is adopted in choosing models at the outset.

9.2.2 Type 2 Indicators (Based on Covariates and Survey Variables)

In this section, we discuss two indicators that are based upon the relationship between the X variables available from the sampling frame and paradata and the survey outcome variables: the fraction-of-missing information and the covariance between survey variables and response propensities. See Wagner (2008, 2010, 2012).

The fraction-of-missing information originates from the multiple imputation literature and reflects the loss of efficiency due to missing data. It is defined as the ratio of the between imputation variance over the sum of the between and within imputation variances. The between imputation variance results from imputing the missing part; the smaller it is, the more certain the model is about choosing the missing values. The within imputation variance results from the uncertainty about the values of population parameters given a finite sample; the smaller it is, the larger the sample and/or the less natural variation there is in the variables. The imputation model predicts each missing value. If the model is based upon regression, then this uncertainty is characterized by drawing values of the regression coefficients from distribution based upon the estimated coefficients and their standard errors and drawing from the residual variance. When multiple draws of this stochastic element are added to the predicted value, this creates a distribution of imputed values. These imputed values will have more variation when the relationship between X and Y is weak.

The indicator is meaningful only when the imputation accounts for the uncertainty in the imputation model. For example, one could impute a constant value for all missing values, for example, the mean of the observed values. This imputation then leads to a zero between imputation variance, but does not account for the variation of the variables around the mean, and gives a false picture of a gain in efficiency.

The population FMI can be defined as

$$\mathrm{FMI}(Y,X)=\frac{(1-\bar{\rho})(1-CD(Y,X))}{\bar{\rho}+(1-\bar{\rho})(1-CD(Y,X))}, \tag{9.4}$$

where $CD(Y,X)$ is the coefficient of determination, that is, the fraction of variance of Y explained by X. When $CD(Y,X) = 0$, that is, when the auxiliary variables do not at all explain the survey variables, then $\mathrm{FMI}(Y,X)=1-\bar{\rho}$. When $CD(Y,X) = 1$, that is, when the auxiliary variables fully explain the survey variables, then $\mathrm{FMI}(Y,X) = 0$. From Equation 9.4, it can be seen that the MI is a measure of efficiency loss and not of bias; even when nonresponse is unrelated to the variable of interest Y, but $CD(Y,X) = 0$, then $\mathrm{FMI}(Y,X)=1-\bar{\rho}$.

A second indicator that employs observed survey variables is based upon the covariance between the survey outcome variables and the estimated response propensities. Under the correct model for the response propensities, this is an estimate of the nonresponse bias. It is defined as

$$\text{NRB}(Y,X)=\frac{\text{cov}(Y,\rho_X)}{\bar{\rho}}. \tag{9.5}$$

As for the type 1 indicators, all estimators need to be estimated given response data and sample data. The coefficient of determination, *CD*(*Y*,*X*), and the covariance between the survey variable and the response propensities, cov(*Y*,*ρX*), can be estimated only for respondents, and will, in general, be biased as an estimate for the whole sample. The estimation of models for survey variables depends on the type of variable. For example, logistic regression may be used for binary survey variables. We refer to the general literature on multivariate analysis, for example, Agresti (2002) and Gelman and Hill (2007).

In general, the overall nonresponse bias indicators can be evaluated for five purposes: (1) to compare the impact of nonresponse over different surveys, (2) to monitor the impact of nonresponse over different waves of the same survey, (3) to monitor the impact of attrition in panels or longitudinal surveys, (4) to monitor the impact of nonresponse during survey data collection, and (5) to compare the impact of nonresponse for different designs of the same survey. In this handbook, we are concerned with purposes three to five. Of course, the outcomes of evaluations under purposes one and two may be motivations to perform evaluations three to five.

EXAMPLE: LFS

The coefficient of determination for the unemployment rate given all three auxiliary variables is relatively low and around 8%. For the two variables age and break-off the coefficient is close to that value and approximately 7%. Table 9.3 shows the fraction of missing information and the estimated nonresponse bias for the unemployment rate. The FMI is slightly lower than the nonresponse rate

TABLE 9.3

The Expected Nonresponse Rate and Various Type 2 Indicators for LFS Example for the Two Models and at the End of Each Phase

	$X_1 \times X_2$			$X_1 \times X_2 \times X_3$	
		F2F		F2F	
	Web	Standard	Extended	Standard	Extended
Nonresponse Rate	0.780	0.560	0.440	0.712	0.565
FMI(*Y*,*X*)	0.773	0.539	0.425	0.693	0.543
NRB(*Y*,*X*)	−0.020	−0.004	−0.003	−0.008	−0.005

and decreases when data collection phases are added. The estimated nonresponse bias is mostly negative and sizeable after the Web first phase. The estimated bias after the third phase is almost equal to that after the second phase.

EXAMPLE: NSFG

The NSFG monitors the FMI or a number of key statistics each day. The key statistics are imputed using the variables described for the *R*-indicator models in the previous section. Each day, 200 imputed values are created for each of the key statistics. From these imputed datasets, the FMI is calculated. Figure 9.4 shows the daily FMI for the mean number of live births reported by female respondents. The upper line with a circle at each point represents the nonresponse rate. The lower line represents the FMI for this variable on a daily basis.

The FMI, in this case, is calculated based on fully imputed datasets—that is, each dataset includes this variable imputed for the entire sample. It is also possible to compare the fully imputed estimate to the estimate based on the respondent data only. The difference indicates that there is differential response across groups of cases defined by the variables used as predictors in the imputation model. Larger differences mean that the imputed values are more different from the observed values and, by extension, the nonrespondents are more different from the respondents when measured by the variables used as predictors in the imputation model.

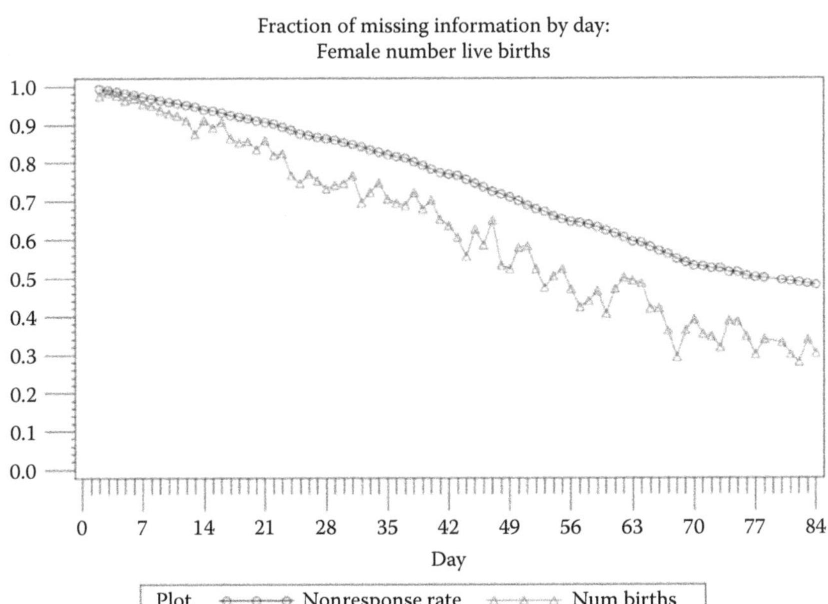

FIGURE 9.4
NSFG fully imputed mean number of life births by female respondents and nonresponse rate during data collection.

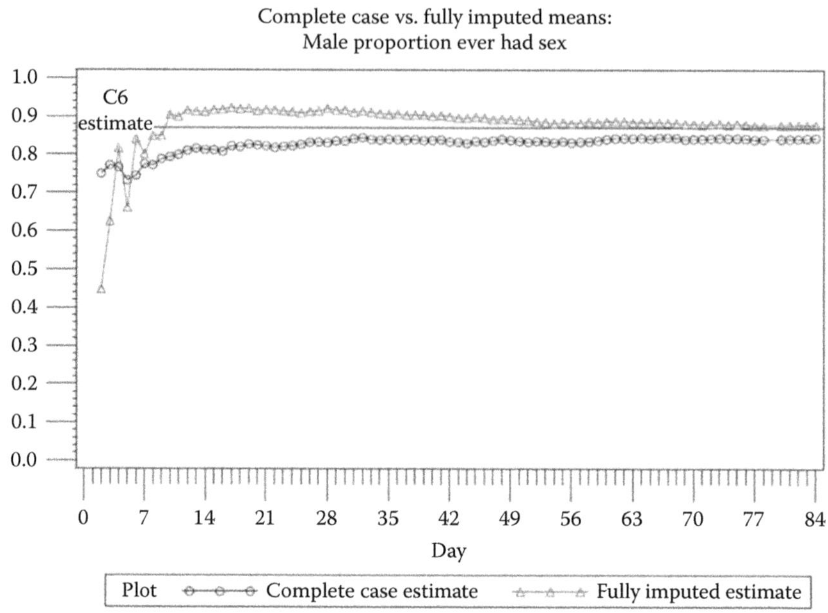

FIGURE 9.5
NSFG fully imputed and respondent proportion of males that have ever had sex.

Figure 9.5 shows the proportion of males that have ever had sex from the fully imputed sample (the line with the triangles), the mean from the respondents (the line with the circles), and a reference line based upon the fully adjusted estimate from a previous cycle.

9.3 Indicators Decomposing the Variance of Response Propensities

So far, the indicators we have examined have been used to give an overall view of the impact of nonresponse on bias. In this section, we discuss indicators that provide more detail and allow for identifying subpopulations that require more attention or, conversely, less attention. Such indicators are usually termed partial or marginal indicators, in a close analogy to partial and marginal correlations.

From monitoring and comparing to intervening and acting is a big step. Chapter 7 deals with this step extensively. However, in order to inform decisions in design and intervention, subpopulations need to be identified that affect nonresponse bias the most. The overall indicators are clearly not fit for that purpose as they provide single values, but the partial or marginal indicators are suited for the identification of subpopulations, or strata, that may require special interventions.

The decision about whether to evaluate strata with respect to nonresponse using the detail available in these partial indicators depends on the magnitude of the overall indicators. When an indicator passes a certain threshold, then the risk of bias may be considered too large and action may be needed. The choice of a threshold depends upon the survey context. The indicators can only be evaluated relative to the same indicator for other surveys. There are two possible strategies for determining a threshold. The first is based on a comparison of different surveys and the second is based on a comparison of the same survey over time. A comparison of indicator values over a wide range of surveys using the same auxiliary information may provide a general threshold to be applied in any specific survey. A historic comparison of indicator values for a single survey may provide a specific threshold to be applied in future waves. The partial or marginal indicators are estimated once the overall indicators do not meet the selected external or internal thresholds.

In the next subsections, we describe the two levels of detail that can be added: the level of an auxiliary variable, and the level of categories of an auxiliary variable. In the following, we will term the indicators partial indicators.

9.3.1 Partial Variable Level

To date, partial indicators have only been developed for type 1 indicators, that is, for indicators based on auxiliary variables from frame data, administrative data, or paradata.

The R-indicator, $R(X)$, and the coefficient of variation of the response propensities, $CV(X)$, are essentially transformations of the variance of response propensities, $S^2(\rho_X)$. There are standard techniques in multivariate analysis for decomposing variances into between and within variances. These are known as the analysis of variance or ANOVA. This approach is used by Schouten, Shlomo, and Skinner (2011) to decompose variances to the variable level. For example, in a logistic regression model with age in five categories and sex in two, the variance of the estimated response propensities would be decomposed into that explained by the age variable and that by the sex variable.

Two types of variable-level partial indicators are distinguished, unconditional and conditional. Unconditional partial indicators measure the impact of single auxiliary variables, but are not adjusted for the other auxiliary variables. Conditional partial indicators adjust the impact of single variables for their associations to the other variables, that is, they adjust for collinearity. Unconditional partial indicators point at variables that give the largest contribution to the variance of response propensities, and conditional partial indicators point at the variables that give the largest specific contribution. When the auxiliary variables in X are uncorrelated, then the conditional values are approximately equal to the unconditional values. When the auxiliary variables in X are fully correlated, then the conditional values are all approximately equal to zero. This feature points at the necessity to be parsimonious in the selection of auxiliary variables. Ideally, the auxiliary variables in X show little association to each other and each measure a specific proportion of response propensity variation. Since such a set of auxiliary variables does not exist, unless artificially constructed by, for instance, a factor analysis model, the conditional partial indicators are intended to adjust for the overlap between the variables.

In the following, we assume that the auxiliary vector X consists of L variables, $X_i = (X_{1,i}, X_{2,i}, \ldots, X_{L,i})^T$.

The unconditional partial R-indicator for variable X_l, denoted as $R_U(X_l)$, with $1 \leq l \leq L$, measures the case-level variability in response rates across the categories of this variable. It is defined as the square root of the between variance of the response propensities over the categories of X_l. $R_U(X_l)$ is defined as

$$R_U(X_l) = \sqrt{\frac{1}{N}\sum_{i=1}^{N}(\rho_{X_l}(x_{l,i}) - \bar{\rho})^2}. \tag{9.6}$$

The conditional partial R-indicator for variable X_l, denoted as $R_C(X_l)$, with $1 \leq l \leq L$, is defined as the square root of the within variance of the response propensities over the categories of X_l. $R_C(X_l)$ equals

$$R_C(X_l) = \sqrt{\frac{1}{N}\sum_{i=1}^{N}(\rho_X(x_i) - \rho_{X_{-l}}(x_{-l,i}))^2}, \tag{9.7}$$

where $x_{-l,i} = (x_{1,i}, \ldots, x_{l-1,i}, x_{l+1,i}, \ldots, x_{L,i})^T$ is the vector of auxiliary variables in which variable l is omitted and $\rho_{X_{-l}}(x_{-l,i})$ is the response propensity for a model with all variables except X_l. In a sense, this indicates the relative importance of this variable in producing variability among the response propensities. Larger values indicate more important contributions to the overall variability of the estimated response propensities.

Whereas the overall R-indicator and unconditional partial R-indicator measure deviations from representative response (i.e., equal response rates across subgroups), the conditional partial R-indicators measure deviations from conditionally representative response. That is, given the impact of a set of predictors on the variability of estimated response rates, does the variable in question add much more variability?

In a close analogy to the overall R-indicator and coefficient of variation, partial coefficients of variation can be defined as partial R-indicators divided by the expected response rate

$$CV_{\mathcal{U}}(X_l) = \frac{R_{\mathcal{U}}(X_l)}{\bar{\rho}}, \quad (9.8)$$

$$CV_{\mathcal{C}}(X_l) = \frac{R_{\mathcal{C}}(X_l)}{\bar{\rho}}. \quad (9.9)$$

For some of the other indicators also partial or marginal versions have been developed. Partial or marginal indicators exist for the imbalance indicator, which follow largely the same decomposition of variances. We refer to Lundquist and Särndal (2013) for details. For type 2 indicators, to date, no partial or marginal versions have been developed, as far as we know. This could potentially be done by decomposing the variances $S^2(Y_X)$ and/or the covariances $\text{cov}(Y,\rho_X)$.

EXAMPLE: LFS

In monitoring the data collection, the question is asked which of the variables affect the representation most over all three phases and over the last two interviewer phases conditionally on being a Web nonrespondent. Partial *CV*s are computed for a model excluding X_3 (interviewer observation) at the end of each phase and for a model including all variables for Web nonrespondents at the end of phases 2 and 3. Table 9.4 shows the results. The unconditional Partial *CV*s are in the first three rows while the conditional partial *CV*s are in the last three rows. It turns out that the auxiliary variables are only weakly related to each other, as the unconditional and conditional values are almost equal. For example, the unconditional partial *CV* for X_1 (an indicator variable for age 15–25) after the Web phase is 0.14 while the conditional partial *CV* for this same variables after the Web phase is 0.17. Web break-off (X_2) has the strongest effect on representation in both settings, but its effect decreases strongly in phases 2 and 3 (as expected). Adaptive strategies may focus on sample units that break-off.

TABLE 9.4

The Partial Coefficients of Variation for LFS Example for the Two Models Calculated at the End of Each Phase

	$X_1 \times X_2$			$X_1 \times X_2 \times X_3$	
		F2F		F2F	
	Web	Standard	Extended	Standard	Extended
$CV_{\mathcal{U}}(X_1)$	0.14	0.19	0.12	0.28	0.15
$CV_{\mathcal{U}}(X_2)$	0.43	0.29	0.21	0.79	0.45
$CV_{\mathcal{U}}(X_3)$	–	–	–	0.09	0.08
$CV_{\mathcal{C}}(X_1)$	0.17	0.18	0.11	0.23	0.12
$CV_{\mathcal{C}}(X_2)$	0.43	0.29	0.20	0.77	0.44
$CV_{\mathcal{C}}(X_3)$	–	–	–	0.09	0.07

9.3.2 Partial Category Level

The variable-level indicators of the previous subsection have category-level analogs. Inspection of the category-level indicators for some of the variables is useful, when the corresponding variable-level values are large (and significantly different from zero at a specified confidence level). Essentially, the category-level indicators are the contributions of each category to the total variance attributed by the variable. For noncategorical auxiliary variables, to date, no analog has been developed yet.

Let k be a category of variable X_l, with $1 \le l \le L$, then the category-level indicators are defined as

$$R_U(X_l,k) = \sqrt{\frac{N_k}{N}}\left(\rho_{X_l}(k) - \bar{\rho}\right), \tag{9.10}$$

$$R_C(X_l,k) = \sqrt{\frac{1}{N}\sum_{i=1}^{N} 1_k(x_{l,i})\left(\rho_X(k) - \rho_{X_{-l}}(x_{-l.i})\right)^2}, \tag{9.11}$$

$$CV_U(X_l,k) = \frac{R_U(X_l,k)}{\bar{\rho}}, \tag{9.12}$$

$$CV_C(X_l,k) = \frac{R_C(X_l,k)}{\bar{\rho}}, \tag{9.13}$$

where N_k is the population size of the stratum belonging to category k and $1_k(x)$ is the 0–1 indicator for $x = k$. The category-level unconditional indicators $R_U(X_l,k)$ received a sign in order to indicate in what direction the representation points; a negative value corresponds to underrepresentation and a positive value to overrepresentation. The category-level conditional indicators $R_C(X_l,k)$ do not have a sign as the representation may be different in each cell formed by the other auxiliary variables. Therefore, the differences are squared so that the sign is ignored. The focus is on the variability due to the addition of the variable in question.

EXAMPLE: LFS

In monitoring the data collection, the question was asked which of the variables affect the representation most over all three phases and over the last two interviewer phases, conditionally on being a Web nonrespondent. In Section 9.3.1, we saw that the indicator for Web break-off had the largest impact over all three phases, and also in the interviewer phases, conditional on being a Web nonrespondent. For this reason, the category-level coefficients of variation for Web break-off are inspected; they are presented in Table 9.5. Only the unconditional values are given; the conditional values

TABLE 9.5

The Category-Level Partial Coefficients of Variation of X_1 = Web Break-off for LFS Example for the Two Models Calculated at the End of Each Phase

		$X_1 \times X_2$		$X_1 \times X_2 \times X_3$	
		F2F		F2F	
	Web	Standard	Extended	Standard	Extended
CV_U (X_1, no breakoff)	0.16	−0.11	−0.08	−0.10	−0.09
CV_U (X_1, breakoff)	−0.39	0.27	0.19	0.20	0.17

are virtually the same. The stratum of persons that breaks off in Web has the largest impact on response. Obviously, this impact is negative in the first phase. In the interviewer phases it is positive.

9.4 Nonresponse Bias

Most nonresponse indicators are indirect measures of nonresponse bias. However, nonresponse bias is a meaningless term until a population parameter and estimator are specified. When a population parameter and estimator are chosen, then the nonresponse bias equals the expected difference of the estimator to the population parameter under both the sampling design and response mechanism. In this more advanced section, we derive expressions for the nonresponse bias of a number of commonly used estimators. We will show in Section 9.5 how indicators link to these biases. Since response probabilities and propensities are key quantities in bias expressions, we start with notation and definitions.

9.4.1 Response Probabilities and Propensities

In this chapter, we adopt a model-assisted approach, that is, we do consider a finite population but construct models to explain nonresponse and to explain key survey variables.

Let the population consist of N units, labeled $i = 1, 2, \ldots, N$. Assume a sample is drawn from the population, and let S_i be the 0–1 indicator for being in the sample for unit i. Each sample unit is then allocated to one of a set of possible strategies, and R_i is the 0–1 indicator for response for unit i. Unless stated otherwise, it is assumed that the response of two different units is independent. This assumption implies that R_i is one and the same random variable under all sampling designs.

The possible strategies form a set, denoted by $\mathcal{A}$. We assume that the data collection is structured into T sequential phases, labeled $t = 1,2,\ldots,T$, so that $\mathcal{A} = \mathcal{A}_1 \times \mathcal{A}_2 \times \ldots \times \mathcal{A}_T$. The phases do not necessarily represent time slots, and they may start at different time points for different sample units. In each phase, an action is allocated to unit i. The action in phase t is denoted as $A_{t,i}$, and it is an element from $\mathcal{A}_t$. The set $\mathcal{A}_t$ may contain one element that represents the choice where no action is taken in phase t. The vector $A_i = (A_{1,i}, A_{2,i}, \ldots, A_{T,i})^T$ represents the full strategy that is allocated to the unit.

The sampling design can be translated to inclusion probabilities for each unit in the population. The so-called first-order inclusion probability for a population unit is the probability that the unit is selected in the sample. In order to optimize sample precision, these probabilities may be different for different units. There are also second and higher order inclusion probabilities that represent probabilities that two or more units are sampled simultaneously. We do not need these probabilities in this chapter, but they are needed to evaluate standard errors of estimators. We denote the first-order inclusion probability as $\pi_i = P[S_i = 1]$ and let $d_i = \pi_i^{-1}$ be the design or inclusion weight. Design weights are needed to construct unbiased estimators.

We assume that an individual population unit has a response probability when applying a specified strategy, $a \in \mathcal{A}$. We denote this probability as $\rho_i(a)$ for unit i, and $\rho_i(a) = P[R_i = 1 | A_i = a]$. The response probability reflects the variation in response that is the result of changing individual and data collection circumstances that the survey cannot control or is not willing to control. The definition of a response probability has been subject to debate over the past decades. The postulation of response probabilities conforms to a so-called random response model, as opposed to a fixed-response model in which population units either respond or not. Here, we adopt a random response model. Individual circumstances that cannot be fully controlled are, for example, the mood of the sample person or the family circumstances as a whole. Data collection circumstances that cannot be fully controlled are, for example, weather conditions, big (inter)national events like sports championships, the mood of interviewers, or the malfunction of Internet connections.

The response probabilities can be decomposed and elaborated per phase, or cumulatively over a series of phases. We let $R_{t,i}$ be the 0–1 indicator for a response in one of the phases up to phase t. We let $\rho_{t,i}(a)$ denote the response probability after phase t given that no response was obtained before phase t, that is, $P[R_{t,i} = 1 | R_{t-1,i} = 0, A_i = a]$. Additionally, we let $\rho_{\leq t,i}(a)$ be the cumulative response probability after phase t, that is, $P[R_{t,i} = 1 | A_i = a]$. As a consequence, $\rho_i(a) = \rho_{\leq T,i}(a)$.

A survey has a range of variables of interest. We represent them as a vector $y_i = (y_{1,i}, y_{2,i}, \ldots, y_{K,i})^T$ and will refer to them as survey variables. The K survey variables may have different measurement levels, that is, nominal, ordinal, or continuous.

Auxiliary to the survey, there may be a vector of variables that can be linked to (part of) the sample or is available at the population level. These are represented as a vector $x_i = (x_{1,i}, x_{2,i}, \ldots, x_{L,i})^T$. We will call them auxiliary variables. There are four types of auxiliary variables (see Chapter 3): (1) administrative data variables that can be linked to the sample, (2) administrative data variables for which population distributions are available, (3) paradata or process data variables that can be linked to (part of) the sample, and (4) survey variables can be linked from previous waves in panels or longitudinal surveys. The second type of variables needs to be included in the survey questionnaire, using the same definition and classification as the administrative data.

Let Z be some arbitrary vector of variables, consisting of survey variables and/or auxiliary variables. Throughout, we will use $\bar{Z}_R$ to denote the design-weighted response mean, $\bar{Z}_S$ to denote the design-weighted sample mean, and $\bar{Z}$ to denote the population mean of Z. The design-weighted response, sample and population variances are represented by $S_R^2(Z)$, $S_S^2(Z)$, and $S^2(Z)$. Design-weighted response, sample and population covariances between two such variables, Z_1 and Z_2, are given by $\text{cov}_R(Z_1,Z_2)$, $\text{cov}_S(Z_1,Z_2)$, and $\text{cov}(Z_1,Z_2)$.

By $\bar{Z}_{\mathfrak{R}}$, $S_{\mathfrak{R}}^2(Z)$ and $\text{cov}_{\mathfrak{R}}(Z_1,Z_2)$, we denote the mean, variance, and covariance for respondents in the total population. They are defined as

$$\bar{Z}_{\mathfrak{R}} = \frac{1}{\bar{\rho}N}\sum_{i=1}^{N} \rho_i Z_i, \tag{9.14}$$

$$S_{\mathfrak{R}}^2(Z) = \frac{1}{\bar{\rho}N}\sum_{i=1}^{N} \rho_i (Z_i - \bar{Z}_{\mathfrak{R}})^2, \tag{9.15}$$

$$\text{cov}_{\mathfrak{R}}(Z_1, Z_2) = \frac{1}{\bar{\rho}N}\sum_{i=1}^{N} \rho_i (Z_{1,i} - \bar{Z}_{1,\mathfrak{R}})(Z_{2,i} - \bar{Z}_{2,\mathfrak{R}}), \tag{9.16}$$

and we will see later that they appear when taking the expectation of unadjusted estimators.

Response probabilities are, generally, unknown and may be substituted or approximated by response propensities. The response propensity is the average response probability over population units with the same specified values on a set of variables. Let $\rho_Y(y; a)$ be the response propensity for population units with a value $Y = y$, given that strategy a is applied. When Y contains one or more continuous variables, then the interpretation of an average response probability can no longer be made. In that case, usually, the interpretation is that of a smooth function and local averaging of response probabilities, that is, models are implicitly introduced for dependence on the continuous variables. The response propensity of unit i follows by filling in its value on the variables and is $\rho_Y(y_i; a) = P[R = 1 | Y_i = y, A = a]$. Note that in moving from response probabilities to response propensities, we change from a pure design-based approach to a model-assisted approach, sometimes termed quasi-randomization. In an analogous way, response propensities can be defined for auxiliary variables, that is, $\rho_X(x_i; a)$. Furthermore, we can again distinguish phase response propensities, $\rho_{t,Y}(y_i; a)$, and cumulative phase response propensities, $\rho_{\leq t,Y}(y_i; a)$.

EXAMPLE: LFS

The design consists of three phases, $T = 3$, where the first phase is a Web invitation, the second phase is a standard face-to-face follow-up and the third phase an extended face-to-face effort. The second and third phases may be conducted for only parts of the sample, that is, $\mathcal{A}_2 = \{\text{F2F follow} - \text{up}, \text{no follow} - \text{up}\}$ and $\mathcal{A}_3 = \{\text{F2F extended, no extension}\}$. The auxiliary vector $X = (X_1, X_2, X_3)^T$ consists of three elements: $X_{1,i}$ is the 0–1 indicator for being between 15 and 25 years of age of person i, $X_{2,i}$ is the 0–1 indicator for a break-off during Web of person i, and $X_{3,i}$ is the 0–1 indicator for a normal dwelling status for person i. The response propensities $\rho_{1,X}(x;\text{ Web})$, $\rho_{\leq 2,X}(x;\text{ Web, F2F})$, and $\rho_{\leq 3,X}(x;\text{ Web, F2F, F2F} - \text{ext})$ are given in Table 9.1.

With the use of response propensities, we can now define nonresponse mechanisms. Three missing-data-mechanisms feature prominently in the statistical literature, for example, Little and Rubin (2002): missing-completely-at-random (MCAR), missing-at-random (MAR), and not-missing at-random (NMAR). Their definitions are not without ambiguity, see Seaman, Galati, Jackson, and Carlin (2013). Here, we will also use the definitions somewhat loosely. Nonresponse is termed missing-completely-at-random for a variable Z, denoted as MCAR(Z), when $\rho_Z(z)$ is constant in z. Nonresponse is missing-at-random for a variable Z given variables X, denoted as MAR($Z|X$), when $\rho_{X,Z}(x,z)$ is constant in z for all values of x. When $\rho_{X,Z}(x,z)$ depends on z for at least one value of x, then nonresponse is termed not-missing-at-random or NMAR(Z|X). Note that MCAR(X,Z) implies MAR(Z|X), but MCAR(Z) does not imply that MAR(Z|X) holds. This may be intuitively contradictory, but is in line with the original literature that introduced the concepts.

In addition to (9.14) through (9.16), we define a propensity-adjusted expected mean, variance, and covariance for respondents in the total population

$$\bar{Z}_{\mathfrak{R}|X} = \frac{1}{N}\sum_{i=1}^{N}\frac{\rho_i}{\rho_X(x_i)}Z_i, \tag{9.17}$$

$$S^2_{\mathfrak{R}|X}(Z) = \frac{1}{N}\sum_{i=1}^{N}\frac{\rho_i}{\rho_X(x_i)}(Z_i - \bar{Z}_{\mathfrak{R}|X})^2, \tag{9.18}$$

$$\text{cov}_{\mathfrak{R}|X}(Z_1, Z_2) = \frac{1}{N}\sum_{i=1}^{N}\frac{\rho_i}{\rho_X(x_i)}(Z_{1,i} - \bar{Z}_{1,\mathfrak{R}|X})(Z_{2,i} - \bar{Z}_{2,\mathfrak{R}|X})), \tag{9.19}$$

which will appear in the expectations of propensity-adjusted estimators. When nonresponse is MAR(Z,X), then these parameters are equal to the corresponding parameters for the full population.

A quantity that will be of interest and that comes closest to the notion of nonresponse error, is the individual contrast of a population unit. It is denoted as Δ_i and, for a given strategy a, defined as

$$\Delta_i(a) := \frac{\rho_{i(a)}}{\bar{\rho}(a)} - \frac{1-\rho_i(a)}{1-\bar{\rho}(a)} = \frac{\rho_i(a)-\bar{\rho}(a)}{\bar{\rho}(a)(1-\bar{\rho}(a))}. \tag{9.20}$$

The contrast reflects the individual contribution to nonresponse bias on response means over the full range of possible variables on a population. In practice, this quantity is unknowable as it usually depends upon unobserved variables. As for response probabilities, there also exists an analog that is limited to specific sets of variables. The contrast for the auxiliary variables, $\Delta_{X,i}$, is defined as

$$\Delta_{X,i}(a) := \frac{\rho_X(x_i,a)}{\bar{\rho}(a)} - \frac{1-\rho_X(x_i,a)}{1-\bar{\rho}(a)} = \frac{\rho_X(x_i,a)-\bar{\rho}(a)}{\bar{\rho}(a)(1-\bar{\rho}(a))}, \tag{9.21}$$

and corresponds, as we will show later, to the individual contribution of unit i to nonresponse bias of the response mean of the auxiliary variables.

In a close analogy to response propensities, let $Y_X(x,a)$ denote the average value of the survey variables for population units with a value $X = x$ when applying strategy a. Again, when one or more auxiliary variables is continuous, a local smoothing is assumed implicitly. Y_X may also be viewed as the projection of the survey variables on the space spanned by the auxiliary variables X. The dependence on the strategy will mostly be ignored until Chapter 11, where we consider the extension to measurement error. $Y_X(x_i,a)$, or simply $Y_{X,i}$, is the individual projection for unit i.

EXAMPLE: LFS

The variable of interest for person i is the 0–1 indicator Y_i for being unemployed. Table 9.1 (presented earlier in the chapter) contains the projected unemployment rate in the eight strata, for example, Y_X(15–25 years, no break − off, dwelling deterioration, Web) = 0.12. Since the nonresponse depends on all variables, the nonresponse to the survey is NMAR($X_1|X_2,X_3$), NMAR($X_2|X_1,X_3$), and NMAR($X_3|X_1,X_2$). In order to motivate the contrasts, we assume that the response is MAR($Y|X_1,X_2,X_3$), so the response propensities for Y are assumed to be constant within the eight strata. Table 9.6 gives the contrasts $\Delta_{X,i}(a)$ in the eight strata after the first, second, and third phase. On average, the contrasts decrease from the first to the third phase, but only mildly. The contrasts of the strata with break-offs all change sign during the interviewer modes and become overrepresented. Young persons that did not break off and live in dwellings with some deterioration also change sign and become underrepresented during data collection.

TABLE 9.6
Contrasts for LFS Example

Age Category	Web Break-off	Dwelling Deterioration	Contrast		
			Web	F2F	F2F Extended
>25	No	Yes	0.78	0.23	0.17
>25	No	No	−0.10	−0.17	−0.24
>25	Yes	Yes	−1.28	1.45	1.39
>25	Yes	No	−1.28	1.24	1.18
15–25	No	Yes	0.19	−0.38	−0.24
15–25	No	No	−0.40	−0.98	−0.84
15–25	Yes	Yes	−1.28	1.24	0.98
15–25	Yes	No	−1.28	1.04	0.78

In the following, we will omit subscripts and/or dependence on strategies, unless it would be unclear to what variables or strategies we are referring.

9.4.2 Bias Approximations of Unadjusted and Adjusted Response Means

The survey sampling and methods literature often restricts attention to population means and totals, although population regression coefficients, covariances, and variances may be of equal importance, or even may be the focal parameters of an analysis. Also the current literature on nonresponse indicators has focused attention mostly on the association to bias of estimators for means and totals. The traditional response rate is an exception; the response rate is a component in the bias of many, if not all, (sensible) estimators and population parameters. We refer to Brick and Jones (2008) for bias expressions for a range of estimators and parameters. Since virtually all indicators take population means and totals as the primary objective, we will do the same here.

We consider four estimators: (1) the design-weighted response mean of a survey variable, denoted as $\bar{Y}_R$, (2) the modified generalized regression estimator (GREG), denoted as $\bar{Y}_{GREG}$, (3) the inverse propensity weighting (IPW) estimator, denoted as $\bar{Y}_{IPW}$, and (4) the double robust (DR) estimator, denoted as $\bar{Y}_{DR}$. The design-weighted response mean is also termed the Horvitz–Thompson or expansion estimator (Horvitz and Thompson, 1952). It does not employ any auxiliary information and adjusts only for unequal sampling inclusion probabilities. The modified GREG estimator (e.g., Bethlehem, 1988) originates from the sampling theory literature for surveys without nonresponse and is modified or applied to surveys with nonresponse. Essentially, the GREG estimator models the survey target variable(s). The IPW estimator (e.g., Hirano, Imbens, and Ridder, 2003) modifies the Horvitz–Thompson estimator by multiplying the sampling design weights by the inverse of response propensities. The IPW estimator essentially models the nonresponse mechanism. There are various versions of the IPW estimator that result from smoothing response propensities or stratifying the population based on response propensities. The double robust estimator (e.g., Bang and Robins, 2005) combines two (working) models, one for the survey variable(s) and one for the nonresponse mechanism, and may be seen as a fusion of the GREG and IPW estimators. The double robust estimator derived its name from the simultaneous modeling; if either one of the models is correctly specified then the estimator is unbiased, that is, it is doubly robust. As for the IPW estimator, many versions of the DR estimator are possible, depending on the functional forms of the models. Vansteelandt and Vermeulen (2015) discuss optimal estimators within the class of DR estimators. For a general discussion on the various estimators see Schafer and Kang (2008).

The estimators are defined as follows:

$$\bar{Y}_R = \frac{1}{\sum_{i=1}^{n} d_i R_i} \sum_{i=1}^{n} d_i R_i Y_i, \tag{9.22}$$

$$\bar{Y}_{IPW} = \frac{1}{N} \sum_{i=1}^{n} \frac{d_i}{\rho_X(x_i)} R_i Y_i, \tag{9.23}$$

$$\bar{Y}_{GREG} = \frac{1}{N} \sum_{i=1}^{n} d_i (R_i Y_i + (1 - R_i) Y_{\tilde{X}}(\tilde{x}_i)), \tag{9.24}$$

$$\bar{Y}_{DR} = \frac{1}{N} \sum_{i=1}^{n} \frac{d_i}{\rho_X(x_i)} (R_i Y_i + (1 - R_i) Y_{\tilde{X}}(\tilde{x}_i)). \tag{9.25}$$

Various choices for $\rho_X(x_i)$ and $Y_{\tilde{X}}(\tilde{x}_i)$ are possible. The standard approach for the GREG estimator is to use linear regression to derive the $Y_X(x_i)$ (i.e., to use a Euclidean distance and vector inner product). In the DR estimator, the choice of auxiliary variables models for the nonresponse and for the survey variables may be different, hence, the distinction between X and $\tilde{X}$.

The bias of the four estimators can be approximated by

$$B(\bar{Y}_R) = \frac{\text{cov}(Y, \rho)}{\bar{\rho}}, \tag{9.26}$$

$$B(\bar{Y}_{IPW}) = \frac{\text{cov}(Y, (\rho / \rho_X))}{\bar{\rho}}, \tag{9.27}$$

$$B(\bar{Y}_{GREG}) = \frac{\mathrm{cov}(Y - Y_{\tilde{X}}, \rho)}{\bar{\rho}}, \tag{9.28}$$

$$B(\bar{Y}_{DR}) = \frac{\mathrm{cov}(Y - Y_{\tilde{X}}, (\rho/\rho_X))}{\bar{\rho}}. \tag{9.29}$$

The approximations in Equations 9.26 through 9.29 hold for large sample sizes n. The terms $Y_i - Y_{\tilde{X}}(\tilde{x}_i)$ may be viewed as residuals for the prediction of survey variables, whereas the $(\rho_i)/\rho_X(x_i)$ may be viewed as residuals for the prediction of response probabilities. Hence, the DR estimator is biased whenever the two types of residuals are correlated. We refer to Schouten, Cobben, Lundquist and Wagner (2016) for details.

Expressions (9.22) through (9.29) hold in general, but they are most intuitive for continuous variables. For a dichotomous variable, the residuals $Y_i - Y_{\tilde{X}}(\tilde{x}_i)$ have no clear interpretation and one would consider misclassification probabilities instead. For a polytomous variable, the bias over the variable itself is meaningless and one would consider bias in the probabilities per category of the survey variable, that is, create dummy variables and measure bias per dummy variable.

9.4.3 Bias Intervals under Not-Missing-at-Random Nonresponse

From expressions (9.26) through (9.29), it can be derived that biases vanish under different nonresponse missing-data-mechanisms. This can be shown using the following:

$$\mathrm{cov}(Y, \rho) = \mathrm{cov}(Y, \rho_Y), \tag{9.30}$$

$$\mathrm{cov}\left(Y, \frac{\rho}{\rho_X}\right) = \mathrm{cov}\left(Y, \frac{\rho_{X,Y}}{\rho_X}\right), \tag{9.31}$$

$$\mathrm{cov}(Y - Y_{\tilde{X}}, \rho) = \mathrm{cov}\left(Y - Y_{\tilde{X}}, \rho_{Y - Y_{\tilde{X}}}\right), \tag{9.32}$$

$$\mathrm{cov}\left(Y - Y_{\tilde{X}}, \frac{\rho}{\rho_X}\right) = \mathrm{cov}\left(Y - Y_{\tilde{X}}, \frac{\rho_{X, Y - Y_{\tilde{X}}}}{\rho_X}\right). \tag{9.33}$$

The equalities (9.30 through 9.33) require some cumbersome manipulation, but they follow intuitively from the properties of projections. The design-weighted response mean is unbiased when nonresponse is MCAR(Y). The IPW estimator is unbiased when nonresponse is MAR($Y|X$). The GREG estimator is unbiased when nonresponse is MCAR($Y - Y_{\tilde{X}}$). Finally, the DR estimator is unbiased when nonresponse is MAR($Y - Y_{\tilde{X}} \mid X$). The GREG and DR estimators are also unbiased when the residuals $Y - Y_{\tilde{X}}$ are zero, that is, when the model perfectly predicts the survey variable. It can be shown that MCAR($Y - Y_{\tilde{X}}$) is equivalent to MAR($Y|\tilde{X}$). and that MAR($Y - Y_{\tilde{X}} \mid X$) reduces to MAR($Y|X$) when $X = \tilde{X}$. Hence, when $X = \tilde{X}$, then all estimators, except the design-weighted response mean, are unbiased under MAR($Y|X$). In the following, we let $X = \tilde{X}$.

The missing-at-random assumption cannot be tested against the not-missing-at-random assumption, and without, an informed model, it needs to be accepted that the assumption does not hold. Under NMAR($Y|X$), all estimators are biased. However, the extent to which they are biased can be bounded under some assumption that will now be detailed. These bounds may be useful for comparisons across design options, with narrower bounds being preferred. They may also be useful for evaluating the relative risk of nonresponse bias.

The bounds are built upon the fact that, for any three variables Z_1,Z_2, and Z_3, it holds that any correlation between two of these random variables can be bounded using the correlations each of these have to the other random variable

$$\mathrm{cor}(Z_1,Z_2) \geq \mathrm{cor}(Z_1,Z_3)\mathrm{cor}(Z_2,Z_3) - \sqrt{1-\mathrm{cor}^2(Z_1,Z_3)}\sqrt{1-\mathrm{cor}^2(Z_2,Z_3)}, \quad (9.34)$$

$$\mathrm{cor}(Z_1,Z_2) \leq \mathrm{cor}(Z_1,Z_3)\mathrm{cor}(Z_2,Z_3) + \sqrt{1-\mathrm{cor}^2(Z_1,Z_3)}\sqrt{1-\mathrm{cor}^2(Z_2,Z_3)}. \quad (9.35)$$

The two bounds (9.34) and (9.35) imply an interval for the correlation between Z_1 and Z_2. The interval is symmetric with width $\sqrt{1-\mathrm{cor}^2(Z_1,Z_3)}\sqrt{1-\mathrm{cor}^2(Z_2,Z_3)}$ and centered around cor(Z_1,Z_3)cor(Z_2,Z_3). The bounds are sharp, that is, for any value, say φ, in the interval there exist Z_1 and Z_2 with the observed cor(Z_1,Z_3) and cor(Z_2,Z_3) and cor(Z_1,Z_2) $= \varphi$.

Since cov(Z_1,Z_2) = cor(Z_1,Z_2)$S(Z_1)S(Z_2)$, it follows that

$$\mathrm{cov}(Z_1,Z_2) \geq \frac{\mathrm{cov}(Z_1,Z_3)}{S(Z_3)}\frac{\mathrm{cov}(Z_2,Z_3)}{S(Z_3)} - \sqrt{S^2(Z_1) - \frac{\mathrm{cov}^2(Z_1,Z_3)}{S^2(Z_3)}})\sqrt{S^2(Z_2) - \frac{\mathrm{cov}^2(Z_2,Z_3)}{S^2(Z_3)}}), \quad (9.36)$$

$$\mathrm{cov}(Z_1,Z_2) \leq \frac{\mathrm{cov}(Z_1,Z_3)}{S(Z_3)}\frac{\mathrm{cov}(Z_2,Z_3)}{S(Z_3)} + \sqrt{S^2(Z_1) - \frac{\mathrm{cov}^2(Z_1,Z_3)}{S^2(Z_3)}})\sqrt{S^2(Z_2) - \frac{\mathrm{cov}^2(Z_2,Z_3)}{S^2(Z_3)}}. \quad (9.37)$$

Before we proceed, we introduce the coefficients of variation and determination, two quantities that appear frequently in the expressions. The coefficient of variation for a variable Z is defined as its standard deviation over its mean

$$CV(Z) = \frac{S(Z)}{\bar{Z}}, \tag{9.38}$$

and the coefficient of determination $CD(Y,X)$ is defined as the ratio of the predicted variance over the actual variance

$$CD(Y,X) = \frac{S^2(Y_X)}{S^2(Y)}. \tag{9.39}$$

The coefficient of determination is usually denoted by R^2, but in order to avoid notational confusion in Section 9.5, we use the notation CD here.

Schouten, Cobben, Lundquist, and Wagner (2016) construct NMAR bias intervals for the four estimators from Equations 9.30 through 9.31 and 9.26 through 9.29. They employ $Z_3 = Y_X$ and $Z_3 = \rho_X$ as the variables to construct lower and upper bounds. In their derivation they use that

$$S^2(Y - Y_X) = S^2(Y) - S^2(Y_X) = S^2(Y)(1 - CD(Y,X)),$$

$$\frac{S^2(\rho/\rho_X)}{\bar{\rho}^2} \cong CV^2(\rho) - CV^2(\rho_X) \quad \text{by Taylor approximation}$$

$$\text{cov}(Y - Y_X, \rho_X) = 0 \quad \text{and} \quad \text{cov}\left(Y_X, \frac{\rho}{\rho_X}\right) = 0.$$

For the design-weighted response mean and the DR estimator, we will show how such intervals can be constructed. The design-weighted response mean is the only estimator with a bias interval that is not centered around zero under MAR($Y|X$). The other estimators center the bias interval around zero, that is, they remove the detectable bias. However, their bias intervals have a width that is not equal to zero.

For the design-weighted response mean, we have $Z_1 = Y$ and $Z_2 = \rho$, and the two choices, $Z_3 = Y_X$ and $Z_3 = \rho_X$, amount to different lower bounds

$$|B(\bar{Y}_R)| \geq \max\left(0, \frac{|\text{cov}(Y,\rho_X)|}{\bar{\rho}} - \sqrt{CV^2(\rho) - CV^2(\rho_X)}\sqrt{S^2(Y)(1-\text{cor}^2(Y,\rho_X))}\right) \tag{9.40}$$

$$|B(\bar{Y}_R)| \geq \max\left(0, \frac{|\text{cov}(\rho,Y_X)|}{\bar{\rho}} - \sqrt{CV^2(\rho)\left(1-\text{cor}^2(\rho,Y_X)\right)}\sqrt{S^2(Y)(1-CD(Y,X))}\right). \tag{9.41}$$

and upper bounds

$$|B(\bar{Y}_R)| \leq \frac{|\text{cov}(Y,\rho_X)|}{\bar{\rho}} + \sqrt{CV^2(\rho) - CV^2(\rho_X)}\sqrt{S^2(Y)(1-\text{cor}^2(Y,\rho_X))}, \tag{9.42}$$

$$|B(\bar{Y}_R)| \leq \frac{|\text{cov}(\rho,Y_X)|}{\bar{\rho}} + \sqrt{CV^2(\rho)(1-\text{cor}^2(\rho,Y_X))}\sqrt{S^2(Y)(1-CD(Y,X))}. \tag{9.43}$$

The lower bounds in Equations 9.40 and 9.41 may be larger than zero, but, in practice, will often be zero, because correlations between auxiliary variables and nonresponse and/or survey variables are usually weak. In general, it cannot be stated which of the two upper bounds (9.42) or (9.43) is smallest, but it can be stated that the adjustment estimators decrease the upper bound and center the intervals around zero.

For the DR estimator, we have $Z_1 = Y - Y_X$ and $Z_2 = (\rho/\rho_X)$. The two choices, $Z_3 = Y_X$ and $Z_3 = \rho_X$, lead to the same lower and upper bounds and the interval is symmetric around zero.

$$|B(\bar{Y}_{DR})| \leq \sqrt{CV^2(\rho) - CV^2(\rho_X)}\sqrt{S^2(Y)(1-CD(Y,X))}. \tag{9.44}$$

We need to stress that all derivations and expressions in this section are mostly conceptual and to motivate the choice of indicators in Section 9.3; they contain quantities that are unknown in practice. Some of the quantities, like $CV(\rho_X)$, can be estimated without being affected by the missing nonresponse data. However, $S^2(Y)$ and $CD(Y,X)$ need to be estimated on the observed respondent data and will be biased, in general. $CV(\rho)$ is a purely conceptual quantity that cannot be estimated at all. These are, however, useful indicators in that any analysis of missing data requires some untestable assumptions. The assumptions underlying these indicators, in some settings, may be relatively mild. Further, these assumptions can also be tested using a sensitivity analysis involving different vectors of X.

9.5 Indicators and Their Relation to Nonresponse Bias

In Section 9.2, we introduced type 1 indicators, based on auxiliary variables only, and type 2 indicators, including also responses to survey variables. Here, we will link these indicators to expressions that were derived for nonresponse bias of commonly used estimators in Section 9.4.

The absolute bias of the design-weighted response mean of a survey variable $\bar{Y}_R$ can be written as

$$|B(\bar{Y}_R)| = \frac{|cov(Y,\rho)|}{\bar{\rho}} = \frac{|cov(Y,\rho_Y)|}{\bar{\rho}}, \quad (9.45)$$

and it is straightforward to show that it can be bounded from above by

$$|B(\bar{Y}_R)| \leq \frac{S(Y)S(\rho)}{\bar{\rho}}. \quad (9.46)$$

Since the standard deviation of the survey variable is a population parameter unrelated to nonresponse, (9.46) shows that the coefficient of variation of the response probabilities is the parameter that needs to be considered

$$CV(\rho) = \frac{S(\rho)}{\bar{\rho}}, \quad (9.47)$$

when the focus is entirely on population means and totals.

Again, there is an analog based on response propensities

$$CV(X) = \frac{S(\rho_X)}{\bar{\rho}}. \quad (9.48)$$

Schouten, Cobben, and Bethlehem (2009) link the response rate and *R*-indicator by setting bounds to $CV(X)$. From Equations 9.1 and 9.48 it can be derived that when $CV(X) \leq \alpha$, then it must hold that

$$R(X) \geq 1 - 2\alpha\bar{\rho}. \quad (9.49)$$

Hence, when the *R*-indicator is plotted against the response rate, then diagonal lines going down from the coordinate (0,1) represent constant coefficients of variation. These plots are called response-representativeness plots, or simply *RR*-plots.

In Equations 9.40 through 9.44, we have demonstrated that these indicators also appear in bias intervals for adjusted response means.

The imbalance, IMB(X), and distance, dist(X), indicators that are introduced by Särndal (2011), relate strongly to R-indicator and coefficient of variation. They are defined as

$$\text{IMB}(X) = S^2(\rho_X), \tag{9.50}$$

$$\text{dist}(X) = S(\Delta_X). \tag{9.51}$$

With some manipulation, it can be shown that

$$\text{IMB}(X) = \bar{\rho}^2 CV^2(X) = \frac{1}{4}(1 - R(X))^2, \tag{9.52}$$

$$\text{dist}(X) = \frac{S(\rho_X)}{\bar{\rho}(1-\bar{\rho})} = \frac{CV(X)}{1-\bar{\rho}} = \frac{1-R(X)}{2\bar{\rho}(1-\bar{\rho})}. \tag{9.53}$$

From Equation 9.53, it can be deduced that $CV(X) = (1-\bar{\rho})\text{dist}(X)$, which corresponds to the well-known form in the literature that absolute bias equals nonresponse rate times absolute contrast.

Let $\hat{\rho}_X$ be an estimated response propensity for the auxiliary variable vector X. The various indicators are estimated as

$$\hat{R}(X) = 1 - 2S_S(\hat{\rho}_X), \tag{9.54}$$

$$\widehat{CV}(X) = \frac{\hat{S}_S(\hat{\rho}_X)}{\bar{R}_S}, \tag{9.55}$$

$$\widehat{\text{IMB}}(X) = \hat{S}_S^2(\hat{\rho}_X), \tag{9.56}$$

$$\widehat{\text{dist}}(X) = \hat{S}_S(\hat{\Delta}_X). \tag{9.57}$$

Different authors have made different choices in modeling response propensities. Särndal (2011) and Särndal and Lundquist (2014a) construct their indicators starting from nonresponse calibration; their approach amounts to linear regression models for response propensities. Schouten, Cobben, and Bethlehem (2009) employ logistic regression models, although bias and standard error approximations in Shlomo, Skinner, and Schouten (2012) are generalized to arbitrary link functions. Furthermore, it is also possible to use different distance functions. For instance, if the Euclidean distance (or L^2 distance) is replaced by the absolute distance (or L^1 distance), then the indicators resemble dissimilarity indices, for example, Agresti (2002).

EXAMPLE: LFS

Table 9.7 now also contains the values for the imbalance and distance indicators for the two models estimated during the three phases. Both indicators show the same pattern as the *R*-indicator for the first model; indicating that contrast is smallest after phase 1. Like the *R*-indicator and *CV* they point at improved representation for the second model when extended interviewer effort is made in phase 3.

In the bias expressions in Equations 9.40 through 9.44, the coefficient of determination appears as an important quantity. This coefficient is part of FMI. As mentioned in Section 9.2, the FMI can be defined as

$$\mathrm{FMI}(Y,X)=\frac{(1-\bar{\rho})(1-CD(Y,X))}{\bar{\rho}+(1-\bar{\rho})(1-CD(Y,X))} \tag{9.58}$$

by dividing all terms over the expected within variance.

Unlike the response propensities, the projections Y_X can only be derived from the responding sample units. This means, that, instead of FMI(*Y*,*X*), one measures $\mathrm{FMI}_{\mathfrak{R}|X}(Y,X)$ which is defined as

$$\mathrm{FMI}_{\mathfrak{R}|X}(Y,X)=\frac{(1-\bar{\rho})(1-CD_{\mathfrak{R}|X}(Y,X))}{\bar{\rho}+(1-\bar{\rho})(1-CD_{\mathfrak{R}|X}(Y,X))} \tag{9.59}$$

with

$$CD_{\mathfrak{R}|X}(Y,X)=\frac{S^2_{\mathfrak{R}|X}(Y_X)}{S^2_{\mathfrak{R}|X}(Y)}. \tag{9.60}$$

TABLE 9.7

Various Type 1 Indicators for LFS Example

		$X_1 \times X_2$		$X_1 \times X_2 \times X_3$	
		F2F		F2F	
	Web	Standard	Extended	Standard	Extended
R(*X*)	0.802	0.694	0.737	0.527	0.593
CV(*X*)	0.454	0.345	0.236	0.822	0.468
IMB(*X*)	0.010	0.023	0.017	0.056	0.041
dist(*X*)	0.581	0.620	0.534	1.154	0.828

In the context of continuous auxiliary variables and survey variables, Andridge and Little (2011) propose a sensitivity analysis by modeling nonresponse through a proxy pattern mixture model in which the contribution of the survey variable is moderated. They suppose that $\rho_i = \rho_{X+\lambda Y}(x_i + \lambda y_i)$, in which parameter λ is fixed (i.e., not estimated) as a value from $[0,\infty)$. When $\lambda = 0$, then the nonresponse is MAR(*Y*,*X*). When $\lambda > 0$, then nonresponse is NMAR(*Y*,*X*); the larger it is, the further away it is from MAR(*Y*,*X*). Andridge and Little (2011) propose to compute $\text{FMI}_{\mathfrak{R}|X}(Y,X)$ for $\lambda \in [0,\infty)$ where $\rho_i = \rho_{X+\lambda Y}(x_i + \lambda y_i)$ in

$$S^2_{\mathfrak{R}|X}(Y_X) = \frac{1}{N}\sum_{i=1}^{N} \frac{\rho_i}{\rho_X(x_i)}(Y_X(x_i) - \bar{Y}_{\mathfrak{R}|X})^2,$$

$$S^2_{\mathfrak{R}|X}(Y) = \frac{1}{N}\sum_{i=1}^{N} \frac{\rho_i}{\rho_X(x_i)}(Y_i - \bar{Y}_{\mathfrak{R}|X})^2,$$

$$\bar{Y}_{\mathfrak{R}|X} = \frac{1}{N}\sum_{i=1}^{N} \frac{\rho_i}{\rho_X(x_i)} Y_i.$$

The resulting interval of FMI values provides a nonresponse bias adjustment of the naïve FMI, following a specific path into the "universe" of not-missing-at-random nonresponse.

EXAMPLE: LFS

The coefficient of determination for the unemployment rate given all three auxiliary variables is relatively low and around 8%. For the two variables age and break-off the coefficient is close to that value and approximately 7%. Table 9.8 repeats some of the values in Table 9.2 and adds some new

TABLE 9.8

The Expected Nonresponse Rate and Various Type 2 Indicators for LFS Example for the Two Models and at the End of Each Phase

	$X_1 \times X_2$			$X_1 \times X_2 \times X_3$	
		F2F		F2F	
	Web	**Standard**	**Extended**	**Standard**	**Extended**
NR rate	0.780	0.560	0.440	0.712	0.565
FMI(*Y*,*X*)	0.773	0.539	0.425	0.693	0.543
$\text{FMI}_{\mathfrak{R}\mid X}(Y,X)$	0.771	0.540	0.425	0.693	0.543
NRB(*Y*,*X*)	−0.020	−0.004	−0.003	−0.008	−0.005
$\text{NRB}_X(Y,X)$	−0.012	−0.008	−0.005	0.013	0.000

indicators. The expanded table shows FMI and the estimated nonresponse bias, without and with adjustment for the response propensities. The FMI is slightly lower than the nonresponse rate and decreases when data collection phases are added. The calibration of the coefficient of determination to the population distribution has a negligible impact on the FMI. The estimated nonresponse bias is mostly negative and sizeable after the Web first phase.

9.6 Summary

In this chapter, we have summarized the indicators that appear most frequently in the literature and we motivated them from expressions of the bias that results from nonresponse to a survey. We end with a number of observations

- The main distinction between indicators lies in the employment of the survey variable outcomes, labeled as type 1 and type 2 indicators. Type 1 indicators may select auxiliary variables based on their theoretical or empirical associations to the main survey variables, but they do not employ survey variables in their computation. Type 2 variables focus on specific survey variables and do employ their outcomes in the survey. Type 1 indicators are only to a small extent model-based, but may be built on auxiliary variables that are irrelevant to the survey. Type 2 indicators are model-based, as the associations to survey variables are known only for respondents, but are much more focused. In surveys with a wide range of variables or in panels, type 2 indicators may be difficult as a range of variables would need to be identified. This may lead to different conclusions across the different selected variables.
- Within the type 1 indicators, the various options for indicators are very similar and can be related to each other conceptually. In practice, at any given point in data collection, the choice of any particular indicator does not have a strong impact on the identification of sample strata that require more effort. However, in the evaluation over time or over surveys, the choice is important because of the standardization using the response rate that is used for some of the indicators.
- Since all nonresponse indicators depend on the choice and/or availability of auxiliary variables, it is imperative that the particular variables that are used are given together with the model that is used. The selection of auxiliary variables depends on the purpose of the evaluation; in this handbook that is adaptation of data collection.

We refer to Chapter 3 for the selection of variables from frame data, administrative data, and paradata.

- As a consequence of the dependence of indicators on auxiliary variables, their values are meaningful only in a relative context. Comparison values are needed in order to make any interpretation. Internal thresholds may be derived from earlier waves or experiments of the same survey. External thresholds may be derived from reference surveys that have the same set of auxiliary variables and that are deemed very accurate.
- We recommend that at least one overall type 1 indicator is used in monitoring with a fixed set of auxiliary variables that are relevant to the survey variables. Partial type 1 indicators should be employed in addition for a detailed look when overall values point at a lower representativeness/balance or have changed to more negative values.
- All indicators are estimated based on survey sample and response data and are subject to bias and imprecision. Both bias and imprecision are sample-size dependent and are typically inversely proportional to the sample size. Consequently, indicator values cannot be given without a specification of the precision. In practice, this means that conclusions about indicator values need to be tempered by an examination, for example, of confidence intervals to avoid "false positive" or other erroneous decisions that are artifacts of small sample sizes.

10

Adaptive Survey Design and Adjustment for Nonresponse

10.1 Introduction

ASD may be viewed as an attempt to perform nonresponse adjustment in both the design and data collection stages, rather than in the estimation stage alone. Adaptive designs allocate data collection resources in such a way that response is more representative or balanced with respect to a selected set of auxiliary variables. In this chapter, we ask the question whether it is beneficial to adjust by design, when adjustment during estimation can also be performed, employing the same variables. Furthermore, we ask whether additional adjustment of an adaptive design is needed after data collection is finished and how this adjustment should be implemented.

In this chapter, we limit our focus to nonresponse, since it is not standard practice to adjust for systematic measurement bias. Questionnaires may be designed to account for random measurement error through the construction of latent constructs underlying multiple survey items. Furthermore, as argued in Chapter 2, adaptive designs may be oriented at any component of survey error, that is, they may attempt to reduce random measurement error. While variance is certainly an important component of the total error, an important motivation for adaptive designs has been the control of bias, even after post-survey adjustments have been applied.

Three criteria will be used in this chapter for evaluating adaptive designs: bias, variance, and feasibility. We refer to Tourangeau et al. (2016) for a discussion.

We start with the third criterion, that of feasibility. From an implementation point of view, ASDs are harder and more complex than nonadaptive designs; they require more advanced case management systems as well as monitoring and analysis tools, timely control of paradata, and linkage to administrative data. These may not be available for all surveys. Chapter 6 discusses these requirements in greater detail. Although post-survey adjustment strategies may involve complex methodology, these are far less demanding in terms of implementation. So from an implementation point of view, it is far easier to postpone adjustment to the estimation stage.

From a variance point of view, ASDs may be beneficial. It is, however, not straightforward to deduce the impact of adaptive designs on standard errors. The focus on making response more balanced ensures that the distributions of auxiliary variables among the respondents are closer to their population distributions. When one would apply a form of weighting, using the auxiliary variables to model survey outcome variables and/or nonresponse, then the weights will have less variation after an adaptive design has been employed. The weight variance reduction may lead to smaller standard errors.

In addition to reductions in weight variation, adaptive designs allow for more flexibility than nonadaptive designs, and, consequently, make more efficient trade-offs between quality and costs. These efficiencies may translate into more completed interviews, under the same quality and cost constraints, thereby reducing estimated standard errors. The reference to the same quality constraints is crucial, as ASDs typically add quality constraints, like low response propensity variation, that make them incomparable to traditional designs. If the sole objective of an adaptive design would be variance reduction under cost constraints, then it would, in general, perform better than nonadaptive designs, for example, see Beaumont et al. (2014). It must be noted that ASDs may include randomization in the allocation of data collection strategies to population strata, for example, by subsampling nonrespondents for follow-up in subsequent phases. Any randomization leads to an increase in standard errors, as noted by Brick (2013), which to some extent counteracts the gain in weight reduction; the weights themselves have larger sampling variation.

The last point of view is bias reduction, which was the original objective of ASDs. This criterion is based on quality control in two ways. First, designs that have smaller detectable bias than other designs, with respect to a relevant set of auxiliary variables, are conjectured to also show less non-detectable bias on survey variables of interest, even after adjustment. In other words, when there is a larger distance to MCAR(X) nonresponse on the auxiliary variables X, then it is expected that there is also a larger distance to MAR(Y,X) nonresponse on survey variables Y given the auxiliary variables. Second, any deviation from the requested quality may be better actively battled during data collection by considering causes and appropriate counter measures rather than doing so passively afterward.

The potential reduction in variance may not be sufficient reason to adopt the more complex logistics and data collection processes of adaptive designs. The potential for bias reduction may be necessary to justify the adoption of these designs. Proof of bias reduction on survey outcome variables is needed, but such proof is difficult or even impossible to obtain, however, since true values are unknown. There are two options to provide evidence of bias reduction. One is to compare the ASD estimates to benchmark estimates obtained in a design with maximal effort, that is, the reduced effort setting.

Another is to treat auxiliary variables as pseudo-survey outcome variables and evaluate their bias. Obviously, it is unknown whether the full effort design has smaller biases and available auxiliary variables may have relatively weak associations with the survey outcome variables. Consequently, there is a legitimate discussion about the benefits of adaptive designs. In this chapter, we present some empirical evidence for the efficacy of adaptive designs to decrease bias and we discuss theoretical reasons why they may be able to do so. This discussion is a continuation of the more statistical sections in Chapter 9. We will ignore bias and sampling error in survey design parameters in this chapter, which lead to additional sources of error, as explained in Chapter 8.

Although ASDs may aim at a reduction of bias and/or variance, nonresponse adjustment afterward will, in general, still be beneficial. We see four reasons for adding a nonresponse adjustment in the estimation stage. First, response may only be partially balanced, despite potential objectives to achieve balance; there may still be bias left on relevant auxiliary variables. Second, the associations between auxiliary variables and survey outcome variables may be estimated more accurately using the full response collected in the survey. Third, the auxiliary variables that are used to form population strata may become outdated and newer, timelier values may become available later. Fourth and final, new auxiliary variables may become available that could not be used at the start or during data collection.

This chapter is written for an audience with some background in statistics and contains technical content. The outline is as follows. In Section 10.2, we first present some theoretical evidence for the bias reduction potential of ASDs. In Section 10.3, we discuss theoretical conditions for bias reduction after nonresponse adjustment. In Section 10.4, we move to inference and discuss nonresponse adjustment in the wake of adaptive designs. We give examples in Section 10.5 and then provide a number of conclusions and recommendations in Section 10.6.

10.2 Empirical Evidence for Bias Reduction After Adjustment

We start with a discussion of empirical evidence for bias reduction. Since the true values of the statistics of interest are, generally, unknown, it is not at all straightforward to find empirical evidence of adaptive design leading to bias reduction after adjustment. We see two options: one option is to do less, that is, in some of the population strata data collection effort is reduced, and, subsequently, the resulting bias relative to the full effort design is measured. Since redesign is often the consequence of survey budget pressure, this situation is sometimes realistic. However, it is only meaningful when the full effort design is considered to be unbiased or to be least biased under all possible designs that satisfy realistic cost constraints. The latter is, of course, unknown, so that it becomes uncertain whether the reduced design is better or worse. Hence, these comparisons attain only a relative, and not absolute character. One may also look at external criteria about the survey outcome variables to motivate comparisons with the benchmark survey, like the percentage of item-nonresponse, the number of reported events, or the size of post-survey adjustments to judge whether the full effort design is least biased. See also the discussion in Chapter 2. However, as with the other situations described, it is impossible to evaluate the validity of these kinds of assumptions.

Another option is to select a subset of the auxiliary variables, to optimize the ASD on these auxiliary variables and to validate on the other variables. The subset used for validation is treated as "pseudo" survey outcome variables. This approach is not realistic for most production settings, as one would normally employ all available auxiliary variables in the design, but the bias on these pseudo-survey variables can be directly measured and compared. In this section, we use the pseudo-survey variable approach as an evaluation tool.

We summarize the findings from Schouten, Cobben, Lundquist, and Wagner (2016) that are based on 14 survey data sets from three countries. Table 10.1 gives the full names and labels of the data sets. The 14 data sets consist of 11 surveys of persons or households, 2 business surveys, and 1 household panel.

In the data sets, four types of comparisons can be distinguished: (1) a comparison of different surveys or different survey designs with the same target population (DE), (2) a comparison of the same survey with repeated administrations at different time periods (WA), (3) an evaluation of a survey during data collection (such as comparisons based on time, phases, or numbers of visits or calls) (EF), and (4) an evaluation of a survey after different processing steps (PS), where "processing steps" refers to stages like obtaining contact, obtaining participation and recruitment for a panel. The type 3 corresponds to different amounts of effort in the same survey.

Per data set, multiple subsets of survey responses are available for these different types of comparisons. To give a few examples: The Dutch Health Survey 2010 (HS) data set contains responses to three designs: single-mode Web, single-mode face-to-face, and a sequential design with Web followed by face-to-face. The Dutch Longitudinal Internet Panel for the Social Sciences (LISS) data set has the 0–1 outcomes for recruitment of a panel respondent, actual registration of a panel respondent, and attrition after 1, 2, and 3 years. The Swedish Living Conditions Survey 2009 (LCS) data set has responses after a standard number of phone calls and after an extended number of phone calls. The US Survey of Consumer Attitudes (SCA) data set considers two waves of the same survey in different years. To all data sets a range of auxiliary variables was linked from administrative data, census geographical data, and/or paradata.

TABLE 10.1

Overview of the Data Sets in Schouten, Cobben, Lundquist, and Wagner (2016)

Label	Survey	Type of Survey	Type of Comparison
HS	Dutch Health Survey 2010	PE	DE
CVS	Dutch Crime Victimisation Survey 2006	PE	DE
HSCVS	Dutch Health Survey and Crime Victimisation Survey 2010	PE	DE
LFS	Dutch Labour Force Survey 2009–2010	PE	WA
SCS	Dutch Survey of Consumer Sentiments 2009	PE	EF
SCSASD	Dutch Survey of Consumer Sentiments Adaptive Design Pilot Study 2009	PE	DE
STSIND	Dutch Short-Term Statistics Survey Manufacturing Industry 2007	BU	EF
STSRET	Dutch Short-Term Statistics Survey Retail Industry 2007	BU	EF
LISS	Dutch Longitudinal Internet Panel for the Social Sciences 2007–2010	PA	PS
LCS	Swedish Living Conditions Survey 2009	PE	EF
PPS	Swedish Party Preference Survey 2012	PE	EF
SCA	USA Survey of Consumer Attitudes 2011–2012	PE	WA
NSFG	USA National Survey of Family Growth 2006–2010	PE	EF
HRS	USA Health and Retirement Survey 2006 and 2008	PE	WA

Note: Per data set the type of survey (BU = business, PE = person, and PA = panel of persons) and the type of comparison made in the data set (DE = different surveys or designs, WA = different waves, EF = different effort in terms of calls, and PS = different data collection processing steps) is given.

The resulting data sets are diverse; they concern different countries and settings, different survey target populations, different survey modes and different sets of available auxiliary variables. As a result, they are ideally suited to evaluate the impact of the balance of auxiliary variables with respect to patterns of response on bias. We search for signs of smaller nonresponse biases in designs that are more balanced.

The strategy to detect whether the conjectured relation between more balance and smaller bias exists, is to estimate biases and imbalance in each of the designs in a data set and search for consistency in how the designs are ranked over multiple variables. In each of the data sets, the available auxiliary variables were used to make these rankings. The available auxiliary variables were first sorted in a random order and were then added one by one as main effects to models for nonresponse. The first variable was added to an "empty" model. The second variable was added as a new main effect to the model with the first variable, the third variable as a new main effect to the model with the first two variables, and so on.

Each design within a data set was ranked twice; once based on its partial *R*-indicator value and once based on its estimated remaining nonresponse bias after adjustment. Under the first option, a design was ranked as best when it had the best representation according to the *R*-indicators. Under the second option, a design was ranked as best when it had the smallest bias (in absolute sense). A design was ranked as weakest when it had the weakest representation or the largest bias.

Under the partial *R*-indicator option, in each step, except for the first, for a new variable, the variable-level conditional partial *R*-indicator was estimated (see Chapter 9). In the first step, the unconditional value was estimated, since the model contained only that variable. The conditional partial *R*-indicator determines the unique contribution of a new auxiliary variable, accounting for the collinearity to the other variables. When all variables are added a sequence of partial *R*-indicator values is computed for each design in the data set. On each step the designs were subsequently ranked based on their partial *R*-indicator values.

The option to rank designs based on bias took a different approach. Before a new variable was entered into the model, it was used as a "pseudo" survey outcome variable. The bias is estimable in this case because the pseudo-survey variable is actually fully observed. An estimate from the full sample can be compared to a nonresponse-adjusted estimate from the responders. A nonresponse adjustment, based upon the model including all the previously entered auxiliary variables was calculated and used to create a nonresponse-adjusted estimate from the responders only of this pseudo-survey variable. That is, the remaining nonresponse bias of the pseudo-survey variable was estimated after adjustment using the variables already in the model. Again the first variable was treated differently; here, since there was no adjustment model, the unadjusted nonresponse bias was estimated. The nonresponse adjustment employed the generalized regression estimator (see Chapter 9). Finally, in each data set and per step the surveys were ranked based on the partial *R*-indicators and based on the estimated remaining nonresponse bias. The result of each ranking of surveys within a data set is again a matrix with as many rows as there are designs and with as many columns as there are auxiliary variables.

From the ranked designs, associations between increases in the partial *R*-indicators and reductions in nonresponse bias on the pseudo-survey variables can be evaluated. The rankings make it possible to test statistically whether improved balance (measured on one set of auxiliary variables) also leads to reductions in nonresponse bias of adjusted estimates (estimated from another set of auxiliary variables).

In order to check consistency, the matrices with ranks were tested against the hypothesis of random rank inversions. This hypothesis holds if, indeed, the amount of nonresponse on one variable is unrelated to the amount of nonresponse on another variable, after adjusting for collinearity, that is, associations between the variables. In case the hypothesis is rejected, that is, when the number of inversions is smaller than would be expected, then this points at a consistency in the performance of designs with respect to partial *R*-indicators and sizes of nonresponse bias. In other words, some designs consistently perform better than others.

Schouten, Cobben, Lundquist, and Wagner (2016) performed tests for each of the 14 data sets but also pooled data sets based on the country and type of comparison that is made. In this handbook, we provide only the outcomes of the tests after pooling data sets. Tables 10.2 and 10.3 contain the observed numbers of inversions and corresponding *p*-values when multiple data sets are combined into one overall rank test based on, respectively, the institute and the type of comparison.

In Table 10.2, three combinations of data sets are made: all nine data sets from Statistics Netherlands, all five data sets from Statistics Sweden and ISR Michigan, and all 14 data sets. In all cases the *p*-values are smaller than 0.05, and with one exception they are smaller than 0.001. The overall test, thus, indicates that the total observed numbers of inversions are much smaller than expected, if design preferences per variable had been random.

In Table 10.3, four combinations of data sets are made: all four data sets under type 1 (different designs or different surveys), all three data sets under type 2 (different waves of a survey), all five data sets under type 3 (different effort in terms of numbers of calls), and both data sets under type 4 (different steps during data collection). Types 3 and 4 lead to small *p*-values for both indicators, while type 2 leads to large *p*-values. For type 2, there is a mixed picture; only the *p*-value for the remaining bias after adjustment is small. However, the number of data sets per type of comparison was relatively

TABLE 10.2

Expected Numbers of Inversions, Observed Numbers of Inversions, and p-Values for Combined Data Sets from Statistics Netherlands, Stat Sweden, ISR Michigan, and All Institutes

	Number of Inversions			*p*-Value	
	Expected	Observed *Pc*	Observed *B*	*Pc*	*B*
Stat Netherlands	189.5	142	97	<0.001	<0.001
Stat Sweden/ISR	118.5	66	97	<0.001	0.02
All	308	208	194	<0.001	<0.001

TABLE 10.3

Expected Numbers of Inversions, Observed Numbers of Inversions, and p-Values for Combined Data Sets from the Four Types of Comparisons

	Number of Inversions			*p*-Value	
	Expected	Observed *Pc*	Observed *B*	*Pc*	*B*
Type 1—designs	34.5	43	15	0.94	<0.001
Type 2—waves	32.5	36	31	0.77	0.38
Type 3—effort	172.5	99	136	<0.001	<0.01
Type 4—steps	52.5	38	18	0.03	<0.001

small, so that we cannot conclude whether associations are absent or that statistical power is simply too low.

The overall conclusion from the various tests is that they provide empirical evidence for the intuition that on average more nonresponse bias on one pseudo-survey variable coincides with more bias on other auxiliary variable or variables, even after accounting for the associations between these auxiliary variables and the pseudo-survey variable. However, the empirical study, obviously, is limited to variables that are available from administrative data or paradata, and these variables have specific features. Overall, the ranks test shows that there is an association between partial *R*-indicators and the bias of adjusted estimates. However, this association, while strong, does not mean that in every case an increase in the partial *R*-indicators translates into a reduction of bias; at the data set level the picture was much more mixed. This may have two causes: first, the bias may be relatively small to begin with, so that there is not much to be gained either from increasing the partial *R*-indicators through design or from employing nonresponse adjustments. Second, available auxiliary variables may associate only modestly to nonresponse behavior, so that detectable biases are small. The next subsection provides some theory and guidance on the observed associations of auxiliary variables with nonresponse and the implications for bias on other variables.

Before we move to a more theoretical motivation, it is important to remark that ASD is sensible from a quality control perspective. Surveys never achieve a 100% response, so that there is an imminent risk of bias. This risk implies that a constant quality improvement is needed. On balance, these efforts, judging from the available evidence, seem to improve the quality of the data with respect to nonresponse bias. Identifying selective factors in the process of nonresponse, and addressing these are important steps in the quality control process. Since different population units have a range of reasons or causes of their nonresponse, such improvement cannot be uniform. That is, there is not a single underlying cause of nonresponse and, therefore, it is unlikely that there will be a single treatment effective for all strata. Administrative data and paradata provide clues as to why units do not respond and assist in choosing effective (follow-up) strategies. Knowing that a full mailbox or stack of newspapers on the doorstep coincides with lower contact rates because no one was at home for some time, that language barriers are larger in big cities, and participation rates to Web surveys vary by age, helps tailoring and designing data collection strategies. The best empirical evidence may actually come from interviewers that have adapted their strategies to observable characteristics of persons or households.

10.3 Theoretical Conditions for Bias Reduction After Adjustment

Empirical evidence for the utility of ASD is crucial, but what are the theoretical conditions for the designs to be effective? To answer this question, we need to consider underlying missing-data-mechanisms (see Chapter 9 for definitions). Since the exact same survey data and linked auxiliary data may result from a continuum of not-missing-at-random mechanisms, the only option to provide theoretical conditions for efficacy of ASD after adjustment is to model the generation of variables themselves. In this section, we follow the approach in Schouten (2015), where variables are drawn from certain variable generating distributions. In other words, we view the available auxiliary variables as drawn in some random way from the universe of potential auxiliary variables.

In Section 9.2, we showed that the bias of standard nonresponse adjustment estimators lies in intervals that are centered around zero and whose widths are proportional to

$$\sqrt{CV^2(\rho)-CV^2(\rho_X)}\sqrt{S^2(Y)(1-CD(Y,X))}. \tag{10.1}$$

Hence, in bias intervals, the coefficients of variation of both the response propensities based upon a set of auxiliary variables ρ_X, $CV(\rho_X)$, and the (true) response probabilities ρ, $CV(\rho)$, determine the room that is left for NMAR nonresponse to affect bias. Suppose variables are selected in some random way from the universe of variables on a population, then what are the expected values of $CV(\rho_X)$ and of the remaining bias of adjusted response means?

Assume that the variables in X are selected at random from all possible variables on a population, then the following can be shown:

Theorem 10.1

If X is selected at random, then $ECV^2(\rho_X)=C\ CV^2(\rho)$, where C is a constant that is independent of X.

In other words, the expected squared coefficient of variation is proportional to the true squared coefficient of variation. The constant C depends on the average associations between variables, and is, generally, much smaller than one.

Schouten (2015) proceeds to model variable generating distributions, so that the constant C can be expressed in a number of properties of the population.

Now from Theorem 10.1, it follows by a Taylor expansion that

$$\begin{aligned} E\sqrt{CV^2(\rho)-CV^2(\rho_X)} &\cong \sqrt{CV^2(\rho)-ECV^2(\rho_X)}+\frac{1}{2}\frac{\mathrm{var}(CV^2(\rho_X))}{(CV^2(\rho)-ECV^2(\rho_X))^{3/2}} \\ &= \sqrt{\frac{1-C}{C}ECV^2(\rho_X)}+\frac{1}{2}\frac{\mathrm{var}(CV^2(\rho_X))}{((1-C/C)ECV^2(\rho_X))^{3/2}}. \end{aligned} \tag{10.2}$$

This result is important, as it shows that, in general, the expected remaining bias is larger whenever the observable $CV(\rho_X)$ is larger for an arbitrary variable. As a consequence, when $CV(\rho_X)$ is measured for two data collection designs for an arbitrary variable, then the design with the smaller value is expected to also have a smaller true coefficient of variation, and, hence, is to be preferred. Importantly, Schouten (2015) shows that this conclusion continues to hold, when one actively attempts to minimize $CV(\rho_X)$ through ASD. Hence, ASD that manages to improve balance on a random set of auxiliary variables, will in expectation improve balance on any other arbitrary variable.

Although Theorem 10.1 may motivate ASD, three remarks need to be made. First, Theorem 10.1 is about one (randomly drawn) variable, whereas, in practice, there are multiple auxiliary variables. It can be shown that any series of variables that is drawn randomly from the universe of variables can be crossed to create a new, single overarching variable that is itself part of the universe of variables containing all the information of the original set of variables. Theorem 10.1 can then be applied directly to the resulting crossed variable to again draw conclusions about bias. Another option is to consider the mean (squared) coefficient of variation. Say $X_1, X_2, \ldots, X_M$ is a series of independently drawn variables, then the estimator

$$CV_M=\frac{1}{M}\sum_{m=1}^{M}CV(\rho_{X_m}) \tag{10.3}$$

can be used to rank designs.

Second, and perhaps somewhat paradoxically to the first remark, the strength of conclusions depends on the number of randomly drawn variables. In the second term of (10.2), the variance of the observed squared coefficient of variation appears. This variance can be large and may mask differences between designs. This variance depends on the number of variables

$$\mathrm{var}(CV_M)=\frac{1}{M}\mathrm{var}(CV^2(\rho_X)), \tag{10.4}$$

and, as usual, on the sampling design in case response propensities are estimated from sample data. Note that the expectation and variance in this section are taken over by the variable generating mechanism and that the sampling design is ignored in the notation.

Third, and most importantly, auxiliary variables are, generally, not drawn at random from the universe of variables. It is more likely that variables are randomly drawn from a subset of the universe of variables on a population. Theorem 10.1, then, no longer holds, but can be replaced by

Theorem 10.2

If X is selected at random from a subset of variables U, then $ECV^2(\rho_X) = C_U CV_B^2(\rho)$, where C_U is a constant that is independent of X but depends on the subset and

$$CV_B^2(\rho) = \frac{S_B^2(\rho)}{\bar{\rho}^2}$$

in which $S_B^2(\rho)$ is the between variance of response probabilities over the clusters of variables formed by U.

Theorem 10.2 implies that variables, drawn at random from a subset of variables, still allow for conclusions about design preferences within that subset. If $CV(\rho_X)$ is smaller for one design than for another design for a randomly drawn variable from a subset, then that design also has smaller expected coefficient of variation for any other arbitrary variable from the same subset. This theorem can be employed in two ways. First, the available auxiliary variables may be seen as draws from a subset of variables. Then the question is what this subset is comprised of and whether it is sufficiently broad to be interesting for the survey at hand. Second, one may actively construct the subset by considering only variables that associate at a minimal level to a variable of interest. On the one hand, paradata variables on the data collection itself are examples where variables may be selected based on their association to response behavior. Interviewer observations, on the other hand, may be selected based on their association to survey variables.

Obviously, the auxiliary variable generating distributions are unknown, so that Theorems 10.1 and 10.2 are mostly conceptual but helpful for thinking about theoretical motivations for the efficacy of ASDs.

10.4 Adjustment of Nonresponse to ASDs

In this closing section of the chapter, we discuss nonresponse adjustment of response data that was obtained through an ASD. Such adjustment will generally be needed; full balance or perfect representation will virtually never be achieved by design, unless strict quotas are applied, or data of over-represented population strata are discarded. However, in many settings additional auxiliary variables will become available after data collection is completed or auxiliary variables will become available in more timely versions. Finally, the associations between auxiliary variables and survey variables can be estimated with the highest precision, once all response data become available.

How then to adjust response data from an ASD? The answer is simple: As usual, using the most relevant auxiliary variables and the most accurate, that is, usually the timeliest versions of these variables. The relevance of auxiliary variables depends on their associations to the key survey variables and to nonresponse. In case of multi-purpose surveys or panels, there are no specific survey variables and relevance may be ascribed more generally. This book is, however, not about general nonresponse adjustment methods and we refer to Kalton and Flores-Cervantes (2003), Särndal and Lundström (2005), and Bethlehem, Cobben, and Schouten (2011) for general references. There is, however, one subtlety in ASD that needs to be accounted for: ASDs may introduce additional variation when strategy allocation probabilities are in between zero and one, that is, when there is randomization in the assignment of sample units to strategies, or when they are based on paradata. Randomization in the assignment of sample units to strategies, for example, subsampling cases for follow-up in a subsequent data collection phase, may reduce both the bias in survey variables and the size of postsurvey adjustment weights. However, the randomness that is introduced, comes at the cost of increased sampling variation in the postsurvey adjustment weights. Paradata variables may introduce randomness as well, especially when they are related to the data collection process itself, for example, the number, timing, and outcomes of calls or visits. The paradata variables are not as stable as more traditional auxiliary variables and may change from one realization of the survey data collection to another. If different strategies are associated with different response behavior or measurement behavior, then the variation in random assignment and in paradata, generally, leads to an increase in variance of population estimators. In this chapter, we ignore measurement error and focus on nonresponse.

We return to the notation of Chapter 9. Let the design consist of T data collection phases and let the vector $A_i = (A_{1,i}, A_{2,i}, \ldots, A_{T,i})^T$ represent the full strategy that is allocated to unit i. Let $p(a|x)$ be the strategy allocation probability for $A_i = a$ given covariates $x_i = x$. Let $1_a(A_i)$ be the 0–1 indicator for the event $\{A_i = a\}$.

We illustrate the impact of strategy allocation variation for a design with two phases, $T = 2$. Suppose $x_i = (x_{0,i}, x_{1,i})^T$, where $x_{0,i}$ are baseline covariates available at the start of data collection and $x_{1,i}$ are covariates linked from paradata recorded during phase 1. Let $q_i(x)$ be the probability density that

$x_{1,i} = x$ is observed and $1_x(x_{1,i})$ be the 0–1 indicator for the event $\{x_{1,i} = x\}$. Since $x_{1,i}$ is not available at the start of data collection, the full strategy is known only after phase 1. When $p(a|x) \in \{0,1\}, \forall a \in \mathcal{A}$, then strategy allocation is fixed and does not add variation itself. However, when $q_i(x) \notin \{0,1\}$, that is, when paradata variables are subject to variation, then variation is added through the strategy allocation.

Consider the design-weighted response mean for a survey variable Y

$$\bar{Y}_R = \frac{\sum_{i=1}^{n} d_i R_i Y_i}{\sum_{i=1}^{n} d_i R_i} = \frac{\sum_{i=1}^{n} d_i \sum_x \sum_a 1_x(x_{1,i}) 1_a(A_i) R_i Y_i}{\sum_{i=1}^{n} d_i \sum_x \sum_a 1_x(x_{1,i}) 1_a(A_i) R_i}. \tag{10.5}$$

Note that $d_i = (1/1 - p(\varnothing|x_i))$, where Ø represents the "empty" action, that is, not being assigned to any strategy. Let $p_s(a|x) = p(a|x)d_i$ be the strategy allocation probability, conditionally on being sampled. The response mean could be weighted by the inverse conditional strategy allocation probabilities

$$\bar{Y}_{R,p} = \frac{\sum_{i=1}^{n} d_i \sum_x \sum_a (1/p_s(a|x)) 1_x(x_{1,i}) 1_a(A_i) R_i Y_i}{\sum_{i=1}^{n} d_i \sum_x \sum_a (1/p_s(a|x)) 1_x(x_{1,i}) 1_a(A_i) R_i}, \tag{10.6}$$

where $1_a(A_i)/p_s(a|x)$ is set to zero when $p_s(a|x) = 0$. Equation 10.6 amounts to a survey design where essentially all strategies are treated equally, for example, a mix of statistics over single survey modes or a mix of statistics over different incentive conditions. Alternatively, one could weight only by the probability that a unit is allocated to a nonempty strategy in phase 2, $d_{S,i}(a_1) = 1 - p_s(\varnothing_2|a_1,x)$, where $\varnothing_2$ corresponds to an empty action in phase 2. The weighted response mean then is

$$\bar{Y}_{R,S} = \frac{\sum_{i=1}^{n} d_i \sum_x \sum_a d_{S,i}(a_1) 1_x(x_{1,i}) 1_a(A_i) R_i Y_i}{\sum_{i=1}^{n} d_i \sum_x \sum_a d_{S,i}(a_1) 1_x(x_{1,i}) 1_a(A_i) R_i}, \tag{10.7}$$

which amounts to a survey design that assumes all units should receive a treatment in each phase and empty actions are viewed as subsampling that needs to be accounted for.

The weighted means (10.5) through (10.7) point us to the bias and variance viewpoints mentioned in the introduction to this chapter. Ignoring the bias perspective, the strategy allocation probabilities may be chosen such that precision is maximized, analogous to standard sampling design for single strategy surveys. ASDs may be seen as extensions to standard sampling designs and (10.6) or (10.7) may simply be more efficient, see Beaumont, Haziza, and Bocci (2014). However, from a bias point of view, the strategy allocation probabilities may have been chosen such that quality and/or cost indicators (see Chapters 4 and 6) are optimized, and undoing these allocations in the weighting of (10.6) or (10.7) would not at all be sensible. To overcome this duality in the choice of strategy allocation probabilities, ideally, one would optimize an overall measure like MSE. However, since assessments of bias are not possible, and, instead, proxy measures for nonresponse bias are used, it is impossible to perform a direct optimization; standard errors and proxy bias measures are on different scales. The only alternative seems to be to construct mathematical optimization problems, like in Schouten, Calinescu, and Luiten (2013), in which constraints are included for bias proxy measures and/or variance, next to cost constraints. In the following, it is assumed that the strategy allocation probabilities were chosen such that they are optimal and no additional weighting as in Equation 10.6 or 10.7 is needed.

Nonetheless, it is important to show that strategy allocation may add variation to estimators. For this reason we consider the variance of (10.5) stemming from three random mechanisms: the outcomes of the paradata variables, the strategy allocation, and the response. Hence, we ignore the sampling design, assuming that the three mechanisms are not dependent on being sampled. First, write (10.5) as $\bar{Y}_R = (V/W)$. A second-order Taylor approximation, see for example Wolter (2007), leads to

$$\mathrm{var}_{q,p,\rho}(\bar{Y}_R) = \frac{E^2_{q,p,\rho}(V)}{E^2_{q,p,\rho}(W)}\left(\frac{\mathrm{var}_{q,p,\rho}(V)}{E^2_{q,p,\rho}(V)} + \frac{\mathrm{var}_{q,p,\rho}(W)}{E^2_{q,p,\rho}(W)} - 2\frac{\mathrm{cov}_{q,p,\rho}(V,W)}{E_{q,p,\rho}(V)E_{q,p,\rho}(W)}\right), \tag{10.8}$$

where the indices refer to the three random mechanisms. We simplify to independent strategy allocation over sample units, that is, each unit is assigned independently of other units. Now, using that $\mathrm{var}_{q,p,\rho}(Z) = E_{q,p}\,\mathrm{var}_{\rho}(Z) + E_q\,\mathrm{var}_p E_{\rho}(Z) + \mathrm{var}_q E_{p,\rho}(Z)$ for any random variable Z, we get

$$\mathrm{E}_{q,p,\rho}(V) = \sum_{i=1}^{n} d_i \sum_x \sum_a q_i(x)p(a|x)\rho_i(a)Y_i, \tag{10.9}$$

$$\mathrm{E}_{q,p,\rho}(W) = \sum_{i=1}^{n} d_i \sum_x \sum_a q_i(x)p(a|x)\rho_i(a), \tag{10.10}$$

$$\begin{aligned}\mathrm{var}_{q,p,\rho}(V) = &\sum_{i=1}^{n} d_i^2 \sum_x \sum_a q_i(x)p(a|x)\rho_i(a)(1-\rho_i(a))Y_i^2 + \\ &+\sum_{i=1}^{n} d_i^2 \sum_x q_i(x)Y_i^2\left(\sum_a p(a|x)\rho_i^2(a) - \sum_{a_1,a_2} p(a_1|x)p(a_2|x)\rho_i(a_1)\rho_i(a_2)\right) \\ &+\sum_{i=1}^{n} d_i^2 \sum_a \rho_i^2(a)Y_i^2\left(\sum_x q_i(x)p^2(a|x) - \sum_{x_1,x_2} q_i(x_1)q_i(x_2)p(a_1|x_1)p(a_2|x_2)\right),\end{aligned} \tag{10.11}$$

$$\begin{aligned}\mathrm{var}_{q,p,\rho}(W) = &\sum_{i=1}^{n} d_i^2 \sum_x \sum_a q_i(x)p(a|x)\rho_i(a)(1-\rho_i(a)) + \\ &+\sum_{i=1}^{n} d_i^2 \sum_x q_i(x)\left(\sum_a p(a|x)\rho_i^2(a) - \sum_{a_1,a_2} p(a_1|x)p(a_2|x)\rho_i(a_1)\rho_i(a_2)\right) \\ &+\sum_{i=1}^{n} d_i^2 \sum_a \rho_i^2(a)\left(\sum_x q_i(x)p^2(a|x) - \sum_{x_1,x_2} q_i(x_1)q_i(x_2)p(a_1|x_1)p(a_2|x_2)\right),\end{aligned} \tag{10.12}$$

$$\begin{aligned}\mathrm{cov}_{q,p,\rho}(V,W) = &\sum_{i=1}^{n} d_i^2 \sum_x \sum_a q_i(x)p(a|x)\rho_i(a)(1-\rho_i(a))Y_i + \\ &+\sum_{i=1}^{n} d_i^2 \sum_x q_i(x)Y_i\left(\sum_a p(a|x)\rho_i^2(a)\right. \\ &\qquad\left. - \sum_{a_1,a_2} p(a_1|x)p(a_2|x)\rho_i(a_1)\rho_i(a_2)\right) \\ &+\sum_{i=1}^{n} d_i^2 \sum_a \rho_i^2(a)Y_i\left(\sum_x q_i(x)p^2(a|x)\right. \\ &\qquad\left. - \sum_{x_1,x_2} q_i(x_1)q_i(x_2)p(a_1|x_1)p(a_2|x_2)\right),\end{aligned} \tag{10.13}$$

which could be combined with (10.8). The important observation from Equations 10.11 through 10.13 is that many terms disappear when $q_i(x) = 1$

for some x and/or $p(a|x) = 1$ for some a. When both $q_i(x) = 1$ for some x and $p(a|x) = 1$ for some a, then (10.11) through (10.13) reduce to

$$\mathrm{var}_{q,p,\rho}(V) = \sum_{i=1}^{n} d_i^2 \rho_i(a_i)(1-\rho_i(a_i))Y_i^2, \tag{10.14}$$

$$\mathrm{var}_{q,p,\rho}(W) = \sum_{i=1}^{n} d_i^2 \rho_i(a_i)(1-\rho_i(a_i)), \tag{10.15}$$

$$\mathrm{var}_{q,p,\rho}(V) = \sum_{i=1}^{n} d_i^2 \rho_i(a_i)(1-\rho_i(a_i))Y_i, \tag{10.16}$$

which are also variance expressions that one would use if one would (erroneously) neglect the randomness in the paradata and strategy allocation.

We considered the variance of weighted response means, which will, generally, be different from, and usually larger than, that of nonresponse adjustment estimators as presented in Chapter 9. However, although general variance expressions will be hard to derive, it will hold that the added variation pertains to these estimators. In performing nonresponse adjustment of data obtained through ASDs, one should, thus, account for variation that may potentially have been added by randomness in paradata variables and/or strategy allocation. Natural options to do so, would be to resort to resampling methods like bootstrap or jackknife, see Wolter (2007), or to use multiple imputation, see Rubin (1987).

10.5 Example

We consider again LFS to illustrate nonresponse adjustment for an ASD. In the example, we employ subsampling, use a paradata variable that becomes available during data collection and adjust on a new variable that becomes available after data collection is finished.

The LFS has two phases; the first phase is online and the second phase is a mix of telephone and face-to-face interviews. Only nonrespondents to the online first phase are allocated to an interviewer. However, nonrespondents are subsampled based on the age of the sample person in three classes (15–24 years, 25–44 years, and 54–65 years) and a binary indicator for break-off that is observed during the online phase. The subsampling probabilities are chosen such that the overall response propensities are equal in expectation for all subgroups. The response propensities for the two phases and the subsampling probabilities are given in Table 10.4.

TABLE 10.4

Stratum Response Propensities for the Two Phases in LFS and Subsampling Probabilities between Phases 1 and 2

Age	Break-Off	Subsampling Probabilities	Response Propensities	
			Phase 1	Phase 2
15–24	No	1.00	0.20	0.40
	Yes	0.74	0	0.70
25–44	No	0.52	0.30	0.60
	Yes	0.65	0	0.80
45–65	No	0.40	0.40	0.50
	Yes	0.65	0	0.80

The population distribution over the three age classes is 15–24 years = 33%, 25–44 years = 33%, and 45–65 years = 34%. Obviously, whether break-off occurs is a random event and is not known at the time the sample is drawn. The break-off probabilities in the three age classes are, respectively, 24%, 15%, and 26%. A simple random sample is drawn, and the sampling probabilities are chosen such that the probabilities for the age category 15–24 years are twice as large as those for the age category 25–44 years and 1.5 as large as those for the age category 45–65 years. The ratios between the sampling probabilities are based on the unemployment rates of the age categories; the 25–44 years group has the lowest rate, whereas the 15–24 years group has the highest rate.

A month after data collection is completed, an additional auxiliary variable, a binary indicator for being employed, becomes available and is linked to the sample. Table 10.5 gives the probabilities per age × break-off category that a person is employed. Table 10.6 shows the response propensities when employment is crossed with age and break-off.

The key survey variable from the LFS is the unemployment rate, which is a mix of wanting to work more, actively seeking a job, and being available to start a new job. Table 10.6 contains the unemployment rates for the 12 sample

TABLE 10.5

Employment Probabilities Per Age and Break-Off Category

Age	Break-Off	Probability Employed
15–24	No	0.33
	Yes	0.33
25–44	No	0.50
	Yes	0.40
45–65	No	0.50
	Yes	0.40

TABLE 10.6

Stratum Response Propensities for the Two Phases in LFS When the Additional Auxiliary Variable, Yes/No Employed, Is Added

Age	Break-Off	Employed	Response Propensities		Unemployment Rate (%)
			Phase 1	Phase 2	
15–24	No	No	0.10	0.35	20
		Yes	0.40	0.55	5
	Yes	No	0	0.70	60
		Yes	0	0.70	15
25–44	No	No	0.25	0.50	15
		Yes	0.35	0.72	4
	Yes	No	0	0.80	15
		Yes	0	0.80	2
45–65	No	No	0.35	0.40	7
		Yes	0.45	0.62	2
	Yes	No	0	0.80	15
		Yes	0	0.80	5

subgroups based on age, break-off, and employment. We assume that, given these three variables, nonresponse is missing-at-random, see Chapter 9.

To illustrate the impact of adjustment, the unadjusted design-weighted response mean is compared to two adjusted estimates, one with only age and one with age × employment as weighting models. We employ post-stratification. We investigate the bias implications of the subsampling that is aimed at balancing age and break-off by considering two options: include and exclude subsampling weights between phases 1 and 2, that is, un-doing the random assignment or not. Hence, in total we have six estimators. For each estimator, we derive the bias (B), standard error (SE), and root-mean-square error (RMSE).

Four random mechanisms are at play: sampling, break-off, subsampling, and nonresponse. We consider the impact of the various mechanisms by adding them one by one. We start by considering the impact of nonresponse on the standard error. We draw a sample of size $n = 8000$, draw break-off, and draw a subsample indicator in case of nonresponse in phase 1. This sample is then fixed and we replicate nonresponse. Table 10.7 contains the average B, SE, and RMSE. Next, we consider the joint impact of nonresponse and subsampling, that is, we fix the sample and break-off but replicate subsampling and nonresponse. The corresponding B, SE, and RMSE are given in Table 10.7. Subsequently, we include break-off in the replication, and, finally, sampling. Results for these two mechanisms are given in Table 10.8.

The results in Tables 10.7 and 10.8 point at the increase in standard error when the various random mechanisms are added. The strongest increase

TABLE 10.7

B, SE, and RMSE for the Six Estimators Accounting for Nonresponse and for Nonresponse Plus Subsampling Variation

	Nonresponse			Nonresponse and Subsampling		
Estimator	**B (%)**	**SE (%)**	**RMSE (%)**	**B (%)**	**SE (%)**	**RMSE (%)**
Unadjusted, no subsampling	1.7	0.33	1.7	1.7	0.36	1.7
Unadjusted, subsampling	3.3	0.30	3.3	3.3	0.34	3.3
Age, no subsampling	1.7	0.30	1.7	1.7	0.32	3.2
Age, subsampling	2.0	0.29	2.0	2.0	0.33	3.3
Age × employed, no subsampling	0.3	0.31	0.4	0.3	0.34	0.5
Age × employed, subsampling	0.6	0.31	0.7	0.6	0.35	0.7

comes from the nonresponse and the sampling mechanisms. The increase in variance due to the randomness of the break-off paradata is modest. It will have a bigger impact when break-off is added as a weighting variable. The biases do not vanish after adjustment, because break-off is related to nonresponse and to the unemployment rate conditional on age and registered employment; the nonresponse is not-missing-at-random for unemployment rate when conditioning only on these two auxiliary variables. The estimators without subsampling weights perform better than with subsampling weights. This is however

TABLE 10.8

B, SE, and RMSE for the Six Estimators Accounting for All Random Mechanisms Except Sampling and for All Random Mechanisms

	All Mechanisms but Sampling			All Mechanisms		
Estimator	**B (%)**	**SE (%)**	**RMSE (%)**	**B (%)**	**SE (%)**	**RMSE (%)**
Unadjusted, no subsampling	1.7	0.36	1.7	1.7	0.51	1.8
Unadjusted, subsampling	3.3	0.34	3.3	3.3	0.47	3.4
Age, no subsampling	1.6	0.32	1.7	1.7	0.48	1.7
Age, subsampling	2.0	0.33	2.0	2.0	0.48	2.1
Age × employed, no subsampling	0.3	0.34	0.4	0.3	0.50	0.6
Age × employed, subsampling	0.6	0.36	0.7	0.6	0.52	0.8

as argued not necessarily the case for all choices of response propensities and unemployment rates. The weighting adjustment with age leads to a negligible adjustment in bias for estimators without subsampling weights. This comes as no surprise as the subsampling is designed to avoid response propensity variation for age and break-off. The additional adjustment with registered employment does remove a considerable part of the bias.

10.6 Summary

We make a number of observations related to nonresponse adjustment of ASDs

- Nonresponse adjustment of data collected through ASDs is, generally, needed given that new auxiliary variables may become available, that auxiliary variables available during data collection may be outdated or noisy, and that perfect calibration during data collection will usually be infeasible, or, due to cost or other constraints, even undesirable.
- ASDs may be viewed from a bias and from a variance perspective. The two perspectives are very different and may lead to different strategy allocation and optimization. The best solution to combining the two viewpoints may be to define mathematical optimization problems that include constraints on proxy measures of bias and/or variance.
- ASDs originate from attempts to reduce the impact of nonresponse on bias and are based on the rationale that a less balanced or representative response with respect to relevant auxiliary variables is a sign of weaker data collection quality. The designs' motivation follows from quality control principles.
- There is empirical evidence that on average balancing of survey response is effective, even after nonresponse adjustment using the same auxiliary information. However, such signals are relatively weak and many variables are needed to detect it.
- Under the assumption that available auxiliary variables are drawn at random from a subset of the universe of possible variables, improved balance on these variables in expectation pertains to other variables not drawn but in the same subset. If one would be able to draw variables entirely at random, then traces of improved balance translate to the full universe of variables.
- Random strategy allocation and paradata that are subject to variation may lead to larger standard errors for estimators that need to be accounted for in the adjustment and when making inferences.

Section V

The Future of Adaptive Survey Design

11

Adaptive Survey Design and Measurement Error

11.1 Introduction

ASDs originate from the desire to reduce the impact of nonresponse on survey statistics by allowing for data collection strategies to be tailored to different, relevant population strata. Some of the design features that are adapted may, however, also impact other surveys errors, and such errors may offset or reduce gains in nonresponse bias. In this chapter, we extend ASDs to measurement error. Although questionnaires have been optimized with respect to design features under consideration, most notably to the survey mode, we do not explicitly discuss tailoring or adaptation of individual survey items, questionnaire modules, or whole questionnaires to relevant population strata. The extended framework is sufficiently general to include a range of questionnaires as possible treatments or actions in ASD. However, we assume that, once allocated, the questionnaire is fixed and is not changed during data collection based on the answering behavior of respondents. Such tailoring or intervention during the actual interview, for example adaptive testing, may improve data quality or reduce editing or interview costs, but is a further complication and beyond the scope of this handbook and chapter.

Measurement error is the result of one or more deficiencies in the answering process, either consciously by lack of motivation, unconsciously by insufficient cognitive ability, or both. The answering process is usually modeled as consisting of four cognitive steps: interpretation, information retrieval, judgment, and reporting of an answer. Interpretation refers to reading or hearing and understanding the question and type of information that is requested. Information retrieval is the collection of information, from memory and/or from external archives or sources, that is needed to answer the question. Judgment concerns the translation of the information that is collected into an answer, for example, by computations or by weighing the information against the definitions and terminology used in the question. Finally, reporting amounts to seeking the answer category that conforms to the answer that came out of the judgment. Each of these processes can lead to measurement error. However, an adaptive design would benefit from identification of the cause of measurement error. For example, a problem with comprehension—interpreting the question—may be most exhibited by the elderly and less educated. A problem with judgment—such as disclosing sensitive behaviors like drug use—may be more problematic among those who use drugs, possibly the younger population. Measurement error analysis and methodology is a vast research area with a very extensive, and still rapidly growing, literature. It is not possible to discuss all facets of measurement error in detail in this handbook. A general reference is Tourangeau, Rips, and Rasinski (2000). Important for our purpose is to distinguish random from systematic measurement errors, to identify design features that are known to affect measurement error, and to identify methodology to quantify measurement error. In this chapter, we present a number of options to include measurement error. Essentially, all previous chapters still apply, except that the approach presented in Chapter 6 needs to be extended so that the optimization includes additional objectives and/or constraints.

A distinction that needs to be made first is between single-purpose and multi-purpose surveys. Single-purpose surveys have only one or a few key variables and ASD may focus directly on the key variable(s) without the need to consider measurement error over all survey variables. Multi-purpose surveys have a large or diverse set of key survey variables, and ASD becomes a multi-dimensional decision problem and measurement errors may be different in size and type for the different survey variables. In this chapter, we explicitly distinguish between the two types of surveys. Another distinction that can be made is between cross-sectional and longitudinal surveys. Contrary to cross-sectional surveys, longitudinal surveys allow for adaptation based on observed measurement behavior; respondents providing lower data quality may be treated differently in later waves or cycles. In person and household surveys, it is not customary to return to respondents when data seem to be subject to measurement error. In business surveys, contacting respondents is mostly for larger businesses. This absence of intervention during interviews or after interviews have been completed, makes ASDs focusing on measurement and nonresponse inherently different from those focusing on nonresponse alone. As a consequence, adaptation to measurement error amounts often to static ASDs.

Measurement error may be random and/or systematic. Random measurement error adds noise to answers and affects only the variation of answers, not their location. Systematic error adds a bias to the answers and affects the location. ASD may focus on reducing one or both types of error. For multi-purpose surveys it, indeed, seems natural to focus on the combined effect of both types of measurement error as different key variables may be affected differently. For single-purpose surveys, it is more natural to start with systematic error, given the original focus on nonresponse bias reduction.

Design features that have the largest impact on both measurement and nonresponse are the survey mode, the type of reporting (self-reporting or proxy reporting), the type of questionnaire (long-form, short-form, and basic-question), and the interviewer. Other design features that are studied often in ASD for nonresponse, for example the calling strategy and the incentive strategy, have not been reported in the literature to have a strong impact on measurement error.

Measurement error is known to be hard to quantify and observe, unless alternative, direct measurements can be made without the mediation of the respondent. Such data are termed record check data or validation data. However, such measurements also may be subject to error. For example, linked administrative data or sampling frame data may itself depend on the behavior of respondents. Two other options that are frequently applied are paradata about answering behavior and latent variable models representing psychometric constructs. Paradata such as response times, eye-tracking, or keystroke log-files may be informative about the motivation and ability of respondents. Informed latent variable models may restrict the correlations between variables in the questionnaire, and may, thus, reveal behavior that does not conform to the model.

The difficulty of quantifying measurement error points to an important decision that needs to be made in survey design, whether it is adaptive or nonadaptive: the choice of a benchmark design for measurement. In the absence of true values, systematic measurement error cannot be determined in an absolute sense. The choice of benchmark design has a large impact on ASDs attempting to reduce systematic measurement error as we will show in this chapter.

In Sections 11.2 and 11.3, we discuss extensions of ASD framework for, respectively, single- and multi-purpose surveys. In both sections, we make use of mathematical optimization models and mathematical programming. These sections, therefore, include technical material necessary to define these methods. The use of such techniques is not imperative and other approaches may be applied. However, to date, the literature on ASDs accounting for multiple errors is still thin, and much more research is needed. In both sections, examples are given. We end with take-away messages in Section 11.4.

11.2 Single-Purpose Surveys

We start by considering surveys with one or a few main variables. At the time of writing of this handbook, we were only aware of research attempting to reduce systematic error in single-purpose surveys. For this reason, we limit our attention to such errors and do not discuss options to reduce random error. We, first, introduce an optimization problem framework and then discuss optimization of ASD.

11.2.1 Framework

For surveys with a single key survey variable, we consider the direct impact of the choice of design features on estimates in the population strata. In the next subsection, we will discuss optimal choices of design features. Before we can turn to optimization, we need to introduce additional notation, need to discuss benchmark designs, and need to revisit the stratification of the population.

Suppose the population is stratified into G groups, labeled as $g = 1,2, \ldots, G$. Furthermore, suppose we consider one data collection phase with A possible actions or strategies, that is, $a \in \mathcal{A} = \{1,2, \ldots, A\}$. Let π_g be the probability that a population unit in stratum g is sampled, and let $p_g(a)$ be the probability that a population unit in stratum g is allocated to action a given that it is sampled. The population size for stratum g is denoted as N_g, the expected sample stratum size is $n_g = N_g \pi_g$, and the total sample size is $n = \sum_{g=1}^{G} n_g$. The relative sizes of stratum g in the population and sample are, respectively, $W_g = N_g/N$ and $w_g = n_g/n$. Let $\rho_g(a)$ be the response propensity for a population unit in stratum g under action a, and let $c_g(a)$ be the costs for a population unit in stratum g associated with action a. Finally, let Y be the survey variable of interest and let $y_g(a)$ be the mean of Y in stratum g under action a. Y may not correspond to a single survey question and may be deduced from multiple survey questions, for example, the unemployment rate, the total income over all jobs, or the number of victimizations in the past year.

We introduce a basic measurement model, see Biemer and Stokes (1991) and Alwin (2007), and let

$$y_i(a) = \mu_{g(i)}(a) + \lambda_{g(i)}(a)\tilde{y}_i(a) + \epsilon_i(a),$$

where $\tilde{y}_i(a)$ is the "true" value for the survey variable for unit i under action a, $\varepsilon_i(a)$ is a random measurement error with zero expectation, $g(i)$ is the stratum to which unit i belongs, and $\mu_{g(i)}(a)$ and $\lambda_{g(i)}$ are systematic stratum and action dependent measurement errors. For example, mode assignments are one type of action that can produce different measurement errors depending upon the assigned mode. We implicitly assume that the survey variable is continuous. For categorical variables, systematic and random measurement errors are usually defined in terms of transition probabilities, for example Biemer (2010), which we will not do here. In this chapter, we focus on systematic measurement errors.

Since the survey is conducted to estimate Y, we must assume that no true values from record check data or validation data are available. If the records were available, there would not be a reason to conduct the survey. As a consequence of there being no validation data, we need to choose a benchmark estimate to assess the impact of the choice of action. Let this be $y_g(BM)$, that is, the mean value of Y in stratum g under the benchmark action. This benchmark is assumed to be the measurement that is least impacted by measurement error. This assumption is typically buttressed by reference to specialized studies of measurement error or by logic. *BM* does not necessarily have to be an element of the possible action set $\mathcal{A}$; it may be an action that is considered to be too expensive to implement, for example, like face-to-face interviews or self-reporting, or too burdensome to conduct, for example, a long questionnaire form with detailed derivation of the survey variable Y. *BM* may also be a counterfactual, hypothetical design, for example, a multi-mode survey design in which answers are given as in one of the modes, or a design in which proxy reporting is allowed but where answers are given as in a self-report. Such counterfactual estimates can only be derived by making assumptions about answering and response behaviors and usually require advanced experimental designs. We define $D_g(a;BM) = y_g(a) - y_g(BM)$ as the "method effect" in stratum g relative to the benchmark action, when action a is applied. The method effect is the combination of the difference in systematic measurement error and the difference in nonresponse bias that has not been accounted for in the nonresponse adjustment. It is assumed that the benchmark mode, although not necessarily error-free, has a smaller bias to the unknown population parameter. For convenience, we will omit the reference to the benchmark in the notation. The term "method effect" differentiates this conceptually from measurement error, which implies the availability of validation data. Since the action (a) can be things other than mode, this method effect is also broader than the term "mode effect."

To this point, we have taken the stratification of the population as given. However, since we consider both nonresponse and measurement, the choice of stratification is now different from that in Chapter 3. There it was stated that strata are created for two goals: they are influential on estimates, and they respond to different actions in the same way (but differently to the same action). Here, the second goal may be rephrased as follows: they produce the same method effect under different actions, but different method effects under the same action. Since the method effect is the net effect of selection biases and measurement biases, the second goal may be detailed a bit further as: strata both respond and answer similarly to different actions, or differences in responding and answering are mitigated for different actions, but not for the same action. This ambiguity in the stratification goals makes it more complex to come up with measures to distinguish influence, as is done in Chapter 3. To avoid such complexity one may take the following simpler approach, which is akin to the fourth option in Section 3.2: (1) create a detailed stratification based on the estimated, anticipated, or known associations between auxiliary variables and the key survey variable and (2) collapse strata when they are found or expected to be homogenous in both response propensities and systematic measurement error. As explained in the introduction, for measurement error it is less natural to adapt during data collection, as one would rarely return to a respondent to check or alter answers. This simpler approach may then be sufficient. However, more research is clearly needed in this area. In Chapter 8, we consider the sensitivity of ASD optimization, which may be yet another reason to collapse strata.

11.2.2 Mathematical Optimization

Since most ASDs have focused on nonresponse, the method effects $D_g(a;BM)$ considered here are new and are added as survey design parameters. In this section, we discuss how they can be included in optimization problems. Quality and cost criteria in traditional optimization problems may be kept or modified using the method effects.

Before we discuss the indicators, we need to remark that in order to solve optimization problems, functions of the allocation probabilities are preferably linear. Linear optimization problems can be solved analytically and do not require numerical approximation. Second best is that functions are nonlinear but convex. Convexity still is tractable and various numerical methods exist to find global optimal points. For nonlinear, nonconvex functions, convergence of optimization is not guaranteed nor is it guaranteed that an optimal point is global; methods may converge to local optimal points.

Since these optimization problems reply upon numeric indicators of quality, we revisit the nonresponse indicators, ignoring the method effects, as introduced in Chapter 9: (expected) response rate, (expected) R-indicator, and (expected) coefficient of variation. These indicators are "expected" in the sense of functioning as constraints or targets for optimization problems.

$$\text{Response rate} \quad \rho = \sum_{g=1}^{G}\sum_{a=1}^{A} W_g p_g(a)\rho_g(a), \tag{11.1}$$

$$RR\text{-Indicator} \quad R = 1 - 2\sqrt{\sum_{g=1}^{G} W_g \left(\sum_{a=1}^{A} p_g(a)\rho_g(a) - \rho\right)^2}, \tag{11.2}$$

$$\text{Coefficient of variation} \quad CV = \frac{\sqrt{\sum_{g=1}^{G} W_g \left(\sum_{s=1}^{S} p_g(a)\rho_g(a) - \rho\right)^2}}{\rho}. \tag{11.3}$$

In our notation, the response propensities in Equations 11.2 and 11.3 follow the same strata as were used for the allocation of strategies. However, the strata for which nonresponse bias is measured may be chosen differently from the strata that are used to allocate strategies. In most settings, however, they will overlap. The relative sample sizes w_g are functions of the sample inclusion probabilities π_g. It is important to remark that both the R-indicator and *CV* are nonlinear and nonconvex functions of the allocation probabilities and, consequently, complicate optimization problems when they are included.

An important quality indicator is the precision, which we operationalize as the expected number of respondents in a stratum

$$\text{Stratum precision} \quad r_g = n_g \sum_{a=1}^{A} p_g(a)\rho_g(a), \tag{11.4}$$

where *N* is the population size. Again, the strata for precision may be different from the strata that are assigned strategies, but such a distinction is avoided to keep notation simple. Other definitions of precision are possible as well, including those that account for clustering or stratification. We avoid these as there is a rich literature on optimal sample design that discusses most of these features. Furthermore, the precision may be considered only at the sample level and defined as the total expected number of respondents over all strata,

$$\text{Overall precision} \quad r = \sum_{g=1}^{G} r_g$$

Next, we move to cost and logistical constraints. The total cost is defined as

$$\text{Cost} \quad C = \sum_{g=1}^{G}\sum_{a=1}^{A} n_g p_g(a) c_g(a). \tag{11.5}$$

For logistical or practical reasons, one may limit the number of changes in actions on some or all of the strata, for example, allow for only one mode-switch or only one switch from self-reporting to proxy reporting. The expected number of switches in a stratum is defined as

$$\text{Stratum action switching} \quad A_g = \sum_{g=1}^{G} p_g(a) m(a), \tag{11.6}$$

where *m*(*a*) is the number of different actions in strategy *a*, which is, obviously, smaller than or equal to the number of phases in the design.

Other indicators may be selected and defined in terms of strategy allocation probabilities, for example, imbalance indicators, fraction of missing information, total number of calls, or total interviewer workload per region. However, some of the indicators are complex functions of the allocation probabilities and may make it hard to solve optimization problems. Other indicators may lead to detailed stratification schemes for the population, for example, a stratification including interviewer regions, and may lead to optimization problems that are not robust to inaccuracy in estimates of survey design parameters, see Chapter 8.

The method effects $D_g(a;BM)$ can be added in various ways. One option is the expected method effect per stratum

$$\text{Stratum method effect} \quad M_g(BM) = \frac{\sum_{a=1}^{A} p_g(a)\rho_g(a)D_g(a;BM)}{\sum_{a=1}^{A} p_g(a)\rho_g(a)}, \tag{11.7}$$

in which method effects per strategy are weighted by their response propensities. As a consequence, Equation 11.7 represents the expected absolute method effect in a stratum not adjusted for nonresponse using other variables than the variables deployed for the stratification. Accounting for such nonresponse adjustment would make it difficult or even impossible to express the method effect explicitly in terms of the strategy allocation probabilities. Instead of the stratum method effect, one may consider the overall expected method effect as an alternative option

$$\text{Overall method effect} \quad M(BM) = \frac{\sum_{g=1}^{G} W_g \sum_{a=1}^{A} p_g(a)\rho_g(a)D_g(a;BM)}{\sum_{g=1}^{G} W_g \sum_{a=1}^{A} p_g(a)\rho_g(a)}. \tag{11.8}$$

The stratum method effect and the overall method effect are nonlinear and nonconvex functions of the strategy allocation probabilities. However, we will show that when they are included as constraints, then they can be rewritten to be linear functions of the allocation probabilities.

A third option is the expected difference in method effects over pairs of strata. Say g_1 and g_2 are two strata, then the difference is defined as

$$\text{Method effect difference} \quad \Delta M_{g_1,g_2}(BM) = M_{g_1}(BM) - M_{g_2}(BM). \tag{11.9}$$

The method effect difference is a natural indicator when it is important to compare survey variables over strata. Again, these strata can be defined separately from strata used for the allocation of strategies. They can be, for example, strata of analytical interest. If such differences are large, then any cross-strata comparison is hampered by nonstructural differences caused by the data collection instrument and may lead to false conclusions. As an overall indicator, one may look at the expected variance of stratum method effects

$$\text{Method effect variance} \quad VM(BM) = \sum_{g=1}^{G} w_g \left(M_g(BM) - \sum_{h=1}^{G} w_h M_h(BM) \right)^2. \tag{11.10}$$

Like the method effect indicators, the differences and variances of method effects are nonlinear, nonconvex functions of the allocation probabilities, but, unlike the method effect indicators, as constraints they cannot be rewritten to linear expressions. This feature implies that optimization problems will become more difficult to handle when the differences and variances of method effects are included.

Now, we turn to optimization problems that include method effects on a key survey variable. One indicator needs to be chosen as the objective function and other indicators may be added as constraints. Additionally, it is necessary to add that $\pi_g \in [0,1]$, $p_g(a) \in [0,1]$, and $\sum_{a=1}^{A} p_g(a) = 1$. If one does not want to vary strategies within strata (i.e., apply the same strategy to all the cases in the stratum), then the second constraint may be replaced by $p_g(a) \in \{0,1\}$. The restriction to one strategy per stratum has the beneficial side-effect that the optimization problem can often be solved by brute force; the number of possible strategy allocations equals A^G.

An example of an optimization problem may be

$$\min_{p_g} C \quad \text{subject to the constraints}$$

$$|M(BM)| \leq \Psi, \quad r_g \geq \Xi_g, \quad \forall g = 1,2,\ldots,G,$$

$$\pi_g \in [0,1],\, p_g(a) \in [0,1] \text{ and } \sum_{a=1}^{A} p_g(a) = 1,$$

where Ψ and the Ξ_g are specified beforehand. The constraint on the absolute total method effect, $|M(BM)| \leq \Psi$, can be made linear.

$$\left| \frac{\sum_{g=1}^{G} W_g \sum_{a=1}^{A} p_g(a)\rho_g(a)D_g(a;BM)}{\sum_{g=1}^{G} W_g \sum_{a=1}^{A} p_g(a)\rho_g(a)} \right| \leq \Psi$$

leads to the two inequality constraints

$$-\Psi \sum_{g=1}^{G} W_g \sum_{a=1}^{A} p_g(a)\rho_g(a) \leq \sum_{g=1}^{G} W_g \sum_{a=1}^{A} p_g(a)\rho_g(a) D_g(a;BM),$$

$$\sum_{g=1}^{G} W_g \sum_{a=1}^{A} p_g(a)\rho_g(a) D_g(a;BM) \leq \Psi \sum_{g=1}^{G} W_g \sum_{a=1}^{A} p_g(a)\rho_g(a),$$

which can be rewritten to

$$\sum_{g=1}^{G} W_g \sum_{a=1}^{A} p_g(a)\rho_g(a)(D_g(a;BM) + \Psi) \geq 0,$$

$$\sum_{g=1}^{G} W_g \sum_{a=1}^{A} p_g(a)\rho_g(a)(D_g(a;BM) - \Psi) \leq 0.$$

Consequently, the optimization problem is fully linear and can be handled in a straightforward way.

An example of a nonlinear optimization problem may be

$$\min_{p_g} CV \quad \text{subject to the constraints}$$

$$\left|\Delta M_{g_1,g_2}(BM)\right| \leq \Psi, \quad \forall g_1, g_2 = 1,2,\ldots,G \quad C \leq B, \quad r_g \geq \Xi_g, \quad \forall g = 1,2,\ldots,G,$$

$$\pi_g \in [0,1],\ p_g(a) \in [0,1] \text{ and } \sum_{a=1}^{A} p_g(a) = 1,$$

where Ψ, B, and the Ξ_g are again specified beforehand. This optimization problem is much harder to solve.

In the next subsection, we will present a real example linked to the Dutch LFS. A few remarks are needed about optimization software and strategies. For fully linear problems, optimization is fairly standard (Bertsima and Tsitsiklis, 1997) and routines are available in SAS and *R*, like `lpsolve` (Galili, 2015) and `PROC OPTMODEL`. For nonlinear problems, for example, Bertsekas (1995), both commercial and noncommercial software are available. In *R*, there are the packages `nloptr` and `Alabama`. In SAS, there is `PROC NLP`. Most optimization routines require a starting value from which a search is initiated, following an iterative procedure. However, convergence is not guaranteed and, if convergence criteria are met, then it may still be a local optimum. We, therefore, recommend using several starting values. At least one starting value should represent a current (nonadaptive) design, so that any convergence, even if it is local, is an improvement. Furthermore, convergence is faster and more likely, when some inductive reasoning is applied to remove sections of the solution space that cannot contain an optimum. Such sections may require too much budget or have a maximum response rate that is below the prespecified threshold. Finally, it is recommended to start with reduced optimization problems. One may omit one or more constraints or change the roles of an objective function and constraint. The resulting optimal values may be used as starting values for the extended optimization problem. In some cases, like the method effect, it may be possible to rewrite a nonlinear constraint or objective function to a linear equivalent. We refer to Valliant, Dever, and Kreuter (2013) for mathematical programming in the context of survey sampling.

11.2.3 Example

We consider the unemployment rate in the Dutch LFS as an example. The LFS is a multi-purpose survey that also provides statistics about the educational level, the occupations, and the type of economic activity connected to employment, but the unemployment rate is considered the most important statistic by its main stakeholders. We borrow elements from the optimization problem described in Calinescu and Schouten (2015).

The first wave of the Dutch LFS was redesigned from a single-mode face-to-face (FTF) design to a multi-mode design with three modes: Web, telephone, and face-to-face. First, a sample of addresses receives an invitation letter to participate online in the LFS. Next, nonresponding households get up to two reminder letters and are then forwarded to either telephone or face-to-face, depending on whether they have a listed phone number and depending on the size of the household. Up to eight persons of 15 years and older are invited to participate. Proxy reporting by a member of the household core is allowed. The redesign from face-to-face to the three-mode design went through an intermediate step in which several single and two-mode designs were run in parallel experimentally. As a consequence, comparisons can be made for five designs:

1. Web only
2. Telephone only
3. Face-to-face only
4. Web → telephone
5. Web → face-to-face

In the example, we consider the single-mode face-to-face design as the benchmark for the evaluation of nonresponse and measurement error of other single- and two-mode designs.

We explore a static ASD in which the population is stratified beforehand based on linked auxiliary data. The construction of the nine strata is discussed in Chapter 3. They are repeated here for convenience:

1. *Registered unemployed*: Households with at least one person registered to an unemployment office (7.5% of the population).
2. *65+ households without employment*: Households with a maximum of three persons of 15 years and older without a registration to an unemployment office, without employment and with at least one person of 65 years or older (19.8% of population).
3. *Young household members without employment*: Households with a maximum of three persons of 15 years and older without a registration to an unemployment office, without employment, with all persons younger than 65 years, and with at least one person between 15 and 26 years of age (2.4% of population).
4. *Non-western without employment*: Households with a maximum of three persons of 15 years and older without a registration to an unemployment office, without employment, with all persons younger than 65 years and older than 26 years of age, and at least one person of non-western ethnicity (1.5% of population).
5. *Western without employment*: Households with a maximum of three persons of 15 years and older without a registration to an unemployment office, without employment, with all persons younger than 65 years and older than 26 years of age and all persons of western ethnicity (11.0% of population).
6. *Young household member and employed*: Households with a maximum of three persons of 15 years and older without a registration to an unemployment office, with at least one employed, with all persons younger than 65 years, and with at least one person between 15 and 26 years of age (15.6% of population).
7. *Non-western and employed*: Households with a maximum of three persons of 15 years and older without a registration to an unemployment office, with at least one employed, with all persons older than 26 years of age, and at least one person of non-western ethnicity (3.9% of population).
8. *Western and employed*: Households with a maximum of three persons of 15 years and older without a registration to an unemployment office, with at least one employed, with all persons older than 26 years of age and all persons of western ethnicity (33.5% of population).
9. *Large households*: Households with more than three persons of 15 years and older without a registration to an unemployment office (4.9% of population).

In order to optimize the design, estimates are needed per stratum for the probability that a household will respond to each of the five strategies, the method effect for the estimated percentage unemployed of all strategies against face-to-face, and the relative costs for each of the strategies. They are presented in Tables 11.1 through 11.3. The method effect of a strategy is the remaining difference to the face-to-face estimate after nonresponse adjustment with the standard LFS weighting model. The LFS weighting model performs a calibration through the generalized regression estimator. As a result the method effect is a mix of mode-specific measurement biases and mode-specific selection biases that are not removed by the weighting model. Table 11.1 shows clear difference in response propensities between strategies and between strata. Table 11.2 shows that method effects can be large for some strata but there is a less clear pattern over modes. This may be partially due to the relatively large standard errors for unemployment rate differences. The costs are given in Table 11.3. The differences in costs are very large between strategies but marginal over strata.

All in all, there is potential for an ASD to trade-off costs against response propensities and method effects.

As an optimization problem, we minimize the absolute overall method effect against constraints on the precision, the costs, and the method effect differences

$$\min_{p_g} |M(F2F)| \quad \text{subject to the constraints}$$

$$\left|\Delta M_{g_1,g_2}(F2F)\right| \leq \Psi, \quad \forall g_1, g_2 = 1,2,\ldots,G \quad C \leq B, \quad r_g \geq \Xi_g, \quad \forall g = 1,2,\ldots,G,$$

TABLE 11.1

Response Probabilities Per Stratum and Strategy

	Stratum								
	1	2	3	4	5	6	7	8	9
Web	23.2%	23.6%	15.5%	10.8%	27.9%	27.7%	17.5%	36.7%	22.4%
	(0.2)	(0.2)	(0.2)	(0.2)	(0.2)	(0.2)	(0.2)	(0.2)	(0.2)
Telephone	20.8%	41.3%	15.2%	8.6%	31.1%	23.8%	14.3%	33.3%	37.5%
	(0.2)	(0.2)	(0.2)	(0.2)	(0.2)	(0.2)	(0.2)	(0.2)	(0.2)
FTF	52.4%	58.3%	51.0%	41.2%	51.2%	54.9%	46.0%	56.8%	61.4%
	(1.0)	(1.0)	(1.2)	(1.0)	(0.8)	(1.0)	(0.8)	(1.0)	(1.0)
Web–Tel	32.8%	48.4%	23.8%	17.5%	42.1%	41.1%	25.8%	52.1%	24.4%
	(0.2)	(0.2)	(0.2)	(0.2)	(0.2)	(0.2)	(0.2)	(0.2)	(0.2)
Web–FTF	49.8%	58.3%	43.4%	36.6%	52.6%	54.7%	44.3%	62.0%	54.2%
	(0.8)	(1.3)	(0.7)	(0.8)	(0.9)	(0.8)	(0.7)	(0.9)	(0.4)

Note: Standard errors in parentheses.

TABLE 11.2

Method Effects for the Unemployment Rate against Single-Mode Face-to-Face Per Stratum and Strategy

	Stratum								
	1	**2**	**3**	**4**	**5**	**6**	**7**	**8**	**9**
Web	1.5	0.0	−2.3	−4.5	0.9	−0.4	−2.2	0.6	−0.4
	(1.3)	(0.0)	(1.4)	(1.8)	(1.9)	(1.3)	(0.8)	(1.0)	(0.9)
Telephone	−0.1	−0.1	−2.3	−4.9	−0.6	−1.0	−0.8	−0.2	−1.2
	(0.7)	(0.2)	(0.9)	(1.2)	(1.1)	(0.9)	(1.2)	(1.4)	(0.6)
FTF	0.0	0.0	0.0	0.0	0.0	0.0	0.0	0.0	0.0
	(0.0)	(0.0)	(0.0)	(0.0)	(0.0)	(0.0)	(0.0)	(0.0)	(0.0)
Web-Tel	0.9	−0.1	−3.7	−1.7	0.5	−0.7	−3.0	0.6	−0.4
	(1.2)	(0.0)	(1.4)	(2.3)	(1.8)	(1.3)	(1.6)	(2.0)	(0.9)
Web-FTF	0.9	0.0	−1.2	−2.0	0.6	−0.3	−1.2	0.4	−0.2
	(1.3)	(0.0)	(1.4)	(1.8)	(1.9)	(1.3)	(1.6)	(2.0)	(0.9)

Note: Standard errors in parentheses.

TABLE 11.3

Costs Per Sample Address Per Each Stratum and Strategy Relative to Face-to-Face

	Stratum								
Strategy	**1**	**2**	**3**	**4**	**5**	**6**	**7**	**8**	**9**
Web	0.03	0.04	0.04	0.04	0.04	0.03	0.03	0.03	0.03
Telephone	0.13	0.17	0.11	0.10	0.15	0.14	0.11	0.16	0.19
FTF	1.00	1.00	1.00	1.00	1.00	1.00	1.00	1.00	1.00
Web–Tel	0.09	0.12	0.10	0.10	0.10	0.09	0.09	0.08	0.07
Web–FTF	0.72	0.71	0.80	0.84	0.73	0.68	0.81	0.62	0.71

$$\pi_g \in [0,1],\ p_g(a) \in [0,1] \text{ and } \sum_{a=1}^{A} p_g(a) = 1,$$

where $\Psi \in \{0.5\%, 1.0\%\}$, B is equal to a budget needed to do a sample of 12,000 addresses in the new three mode design 5% or the same budget increased by 5%, and the Ξ_g are taken equal to the average number of respondents under the new three mode design. Table 11.4 shows the minimal method effects under different method effect difference and budget constraints. The method effects in strata 3, 4, and 7 are relatively large and the method effect difference constraint forces these strata to be allocated to a mix of FTF and Web–FTF. Since these are small strata, this can be afforded under the budget. Strata 1 and 8 are assigned to telephone, while stratum 2 is assigned to Web. The strata 5, 6, and 9 have mixed strategies. The minimal absolute method

TABLE 11.4
Minimal Absolute Overall Method Effects Given Constraints on the Budget Level and the Stratum Method Effect Differences

Budget Level	Method Effect Difference (%)	Optimal Method Effect (%)
Equal	1.0	0.1
	0.5	0.1
+5%	1.0	0.0
	0.5	0.1

effect can only be made smaller with a 5% increase of the budget level and under the least stringent method effect difference of 1%. When the budget is increased, then sample cases in strata 5, 6, and 9 are partially moved to FTF.

11.3 Multi-Purpose Surveys and Panels

Like single-purpose surveys, the literature about reduction of measurement error in multi-purpose surveys and panels is limited, but is beginning to emerge. Especially, in panels there may be much to gain from ASDs considering measurement behavior and attrition. We begin by summarizing measurement error on multiple survey items through so-called response quality indicators. Next, we proceed to quality and cost functions employing such propensities. Finally, we discuss optimization.

11.3.1 Response Quality Indicators and Propensities

For surveys with many or diverse variables of interest, the approach in Section 11.2.3 may be extended, but often leads to either intractable or less meaningful optimization problems. A method effect constraint may be added per variable. Alternatively, the method effect may be computed as the mean, possibly weighted, of the absolute method effects per variable and added as objective function or as constraint. However, while such an approach may be feasible in the simplest of problems, for most choices of constraints and objective functions the optimization problems become very hard to handle due to their nonlinear forms. A simpler option would be to take the average method effect over a representative sample of survey variables and add this average method effect as constraint or objective function. This is akin to solutions that are sometimes proposed for multivariate sampling problems. The resulting optimization problems have the same complexity as those for individual survey variables, but lead to less sensible designs; much of the information is lost in averaging effects and individual variables may still show great differences. Calinescu and Schouten (2016), therefore, propose to reduce the dimensionality of the survey variables by considering, instead, more general response quality indicators. These response quality indicators are proxy indicators for measurement error. They have the advantage of being measurable. The corresponding propensities of such indicators for different population strata are then included in the constraints and/or objective functions.

A response quality indicator, also termed data quality indicator, summarizes respondent answering behavior that is a sign of an increased risk of measurement error for at least a subset of the survey variables. For example, answering questions more quickly than average or offering "don't know" responses more frequently than average are proxy indicators of poor response quality. It goes beyond the scope of this handbook to give a detailed account of the sources of measurement error and to translate them to an extensive and complete list of observable answering behaviors, for example, see Tourangeau, Rips, and Rasinski (2000) and Baumgartner and Steenkamp (2001). Table 11.5 provides examples of response quality indicators. Three types of indicators are distinguished, indicators based on gold standard data, latent variable models, and answering behavior paradata.

TABLE 11.5

Examples of Response Quality Indicators Based on Gold Standard Data, Latent Variable Models, and Paradata on Answering Behavior

Type	Response Quality Indicator
Gold standard data	Difference to validation data
	Difference to audit or record check data
Latent variable models	Amount of random measurement error in scale items (reliability)
	Loading on common factors/classes representing response styles
Answering behavior	Average duration per completed item
	Variance in durations over completed items
	Average decrease in duration per completed item over course of interview
	Percentage of items with missing data
	Percentage of items with do-not-know answers
	Percentage of items with order effects
	Percentage of items with agree answers
	Occurrence of rounding of answers (continuous measurement levels)
	Percentage of items with answers in nonsensitive categories
	Percentage of items with answers that skip filter questions
	Variance of responses to batteries of items

The three types of response quality indicators in Table 11.5 represent different forms of auxiliary information that can be employed to deduce an increased risk of measurement error. Gold standard data are linked data for the same respondents from another source, often from administrative data. These linked data are anticipated to have good, or even perfect, data quality, and to reveal measurement error explicitly. However, even being perfect measurements, the linked data may refer to another time point than the survey or may use alternative definitions for population units or variables of interest. In general, it is, therefore, hard to come by perfect gold standard data, which makes this option relatively rare in practice. Latent variable models incorporate auxiliary information by constraining sizes and signs of the covariances between variables using informed models. The literature about such models uses the term survey item to describe the combination of a survey question, corresponding answering categories, and any accompanying text or introduction. Latent variable models usually cluster survey items that are strongly related by questionnaire construction. The models may be strengthened by randomizing the order in which survey items appear in the questionnaire or in questionnaire blocks, or by randomizing the direction of answering scales (from negative to positive or vice versa). There is a literature on models for various behaviors, generally termed response styles, in batteries of survey items, for example, Van Vaerenbergh and Thomas (2012) and Bais et al. (in press). A simple example is the inclusion of a constant latent variable for straightlining in batteries of survey items. Latent variable models can only be used for survey questionnaires that, by design, have multiple survey items on a smaller amount of inferred topics or variables (termed latent constructs). The answering behavior indicators include paradata about the answering behavior through time stamps, audit trails, or log files, or even through extensive observation data like eye-tracking. Paradata are by-products of the data collection process or can be collected in a straightforward way using existing data collection channels. In order to be suitable to detect increased risk of measurement error, paradata for measurement error are based on expert knowledge and literature, for example, the average interview duration per survey question that was posed to the respondent or the fraction of answers that skipped follow-up questions after a filter question. The three types of response quality indicators may, of course, be employed simultaneously, but there is very little research in this area. However, for multiple answering behavior indicators, evaluations of data quality often include composite indicators that summarize the individual indicators, for example, at least one response quality indicator was nonzero.

11.3.2 Quality and Cost Functions Based on Response Quality Propensities

We follow the notation of Section 11.2.1 to include propensities for response quality indicators for population strata.

Suppose in the comparison, L response quality indicators, labeled l=1, 2, ..., L, are employed. A threshold is set for each indicator and if the expected value of the indicator lies outside the threshold for a stratum then that stratum is considered to have too much risk of measurement error. For example, intervals may be constructed for the distance of a survey variable to an administrative data variable, for the loading on a constant factor in a latent variable model, or for the average interview time over the survey items that are asked to the respondent. This might be implemented by estimating the propensity that a case in a stratum exhibits the observable characteristic, for example, the propensity of responding to a question too quickly, using data from previous waves or rounds of the survey. Let $Q_g = (Q_{1,g}, \ldots , Q_{L,g})' \in \{0,1\}^L$ be the vector containing the 0–1 values per indicator, that is, the 0–1 dummy variables for attaining values in the critical intervals. Furthermore, let $\theta_g(q_1, \ldots , q_L; a) = P[Q_{1,g} = q_1, \ldots , Q_{L,g} = q_L; a]$ be the joint probability distribution for stratum g, given strategy a is applied. $\theta_g(0, \ldots , 0; a)$ represents the propensity that none of the response quality indicators attains a critical value, while $1-\theta_g(0, \ldots , 0; a)$ is the propensity that at least one response quality indicator falls outside the specified intervals. We call $\theta_g(q_1, \ldots , q_L; a)$ the joint response quality distribution.

From the joint response quality distribution we can define response quality propensities. The marginal response quality distribution of a single indicator is denoted as $\theta_{g,l}(q_l; a)$ and may be viewed as the response quality propensity for indicator l. For multiple indicators, there is a multitude of options to define response quality propensities. Here, we will use $\theta_g^J(a) = 1-\theta_g(0,\ldots,0;a)$ as the joint response quality propensity for strategy a in stratum g, or simply the response quality propensity for stratum g. So we consider the propensity of at least one response quality indicator to have a critical value. However, other choices are possible based on the joint response quality distribution. One may, for instance, weigh the response quality indicators differently in their impact or risk.

In the next subsection, we will include individual or joint response quality propensities in mathematical optimization problems. Before we do, we make three remarks. First, stratum response quality propensities need to be estimated or constructed from experience. Apart from the modeling of such propensities, which may follow approaches as in Chapter 5, the survey data and paradata must of course support such estimation or judgment. In general, this will imply that in the design of the survey questionnaire and data collection, extra effort may be needed, for example, by adding timestamps or randomization in the questionnaire, or by labeling and coding survey items. Second, response quality propensities point to an increased risk of measurement error, and not at actual measurement error. Third, since the propensities are estimated on respondent data, like response propensities they are subject to imprecision. Small samples do not allow for strong conclusions and the optimization should be robust in order to account for the imprecision, see Chapter 8.

11.3.3 Mathematical Optimization

We see two main approaches to include response quality propensities into ASD optimization problems: (1) modify the response rate by the response quality rate or (2) constrain the response quality rate.

The modified response rate multiplies the response propensity by the response quality propensity. Doing so, it corresponds to the response rate for respondents without increased risk of measurement error

$$\text{Modified response rate} \quad \rho_{RQ} = \sum_{g=1}^{G}\sum_{a=1}^{A} W_g p_g(a)\rho_g(a)\theta_g^J(a). \tag{11.11}$$

The alternative option is to constrain the response quality rates for each stratum or over all strata. The response quality rate in stratum g is defined as

$$\text{Stratum response quality rate} \quad rq_g = \frac{\sum_{a=1}^{A} p_g(a)\rho_g(a)\theta_g^J(a)}{\sum_{a=1}^{A} p_g(a)\rho_g(a)} \tag{11.12}$$

and the overall response quality rate is

$$\text{Response quality rate} \quad rq = \frac{\sum_{g=1}^{G} W_g \sum_{a=1}^{A} p_g(a)\rho_g(a)\theta_g^J(a)}{\sum_{g=1}^{G} W_g \sum_{a=1}^{A} p_g(a)\rho_g(a)}. \tag{11.13}$$

Hence, the response quality rate is the modified response rate divided over the unmodified response rate. Both Equations 11.12 and 11.13 can be rewritten such that they become linear in the strategy allocation probabilities. This can be done analogously to the method effect constraints in Section 11.2.2.

An example of an optimization problem that maximizes the modified response rate is given by

$$\max_{p_g} \rho_{RQ} \quad \text{subject to the constraints}$$

$$R \geq \Gamma, \quad C \leq B, \quad r_g \geq \Xi_g, \quad \forall g = 1,2,\ldots,G,$$

$$\pi_g \in [0,1],\ p_g(a) \in [0,1], \text{ and } \sum_{a=1}^{A} p_g(a) = 1$$

where Γ, B, and Ξ_g are specified. In other words, the modified response rate is maximized given a minimal required R-indicator, given a maximal budget, and given minimal numbers of respondents per stratum.

An example of an optimization problem that constrains the stratum response quality rates is given by

$$\max_{p_g} R \quad \text{subject to the constraints}$$

$$rq_g \geq \Gamma_g, \quad C \leq B, \quad r_g \geq \Xi_g, \quad \forall g = 1,2,\ldots,G,$$

$$\pi_g \in [0,1], \, p_g(a) \in [0,1], \sum_{a=1}^{A} p_g(a) = 1,$$

where Γ_g, B, and Ξ_g are specified. This problem maximizes the R-indicator given maximal stratum response quality rates, given a maximal budget, and given minimal numbers of respondents per stratum.

The same guidelines and software as in Section 11.2.3, can be used for the optimization problems of this section. The two main approaches, modify the response rate and constrain response quality rates, are very different in nature and, generally, lead to different optimal designs. The two approaches have similar tractability in terms of optimization properties, as the modified response rate is linear in the strategy allocation probabilities and the response quality rate constraint can be rewritten to a linear constraint. However, they do have different conceptual meanings; a modified response rate implies that respondents who show measurement risk are discarded, while a response quality rate implies that lower quality responses are to be avoided but not discarded.

11.3.4 Example

Again, we consider the Dutch LFS as the example. In this section, we take the viewpoint of other LFS users and view the LFS as a multi-purpose survey that produces statistics about the full employment and educational status of the labor force population. Here, we borrow elements from Calinescu and Schouten (2016).

Since the LFS is a household survey, the choice to allow for proxy reporting by one member of the household has both cost implications relating to nonresponse and measurement issues. Proxy reporting increases contact propensities, and, as a consequence, also improves response propensities and lowers costs per sample unit. Since proxy reporters can be less knowledgeable, there is a risk of increased measurement error. For this reason, we consider the type of reporting, self-reporting only versus proxy reporting as two strategies in ASD.

We consider an ASD in which the population is stratified beforehand and adaptation takes place at the outset. We stratify the population based on the average age of the household members between 15 and 65 years of age (the labor force population). We consider three strata 15–25 years, 25–54 years, and 55–65 years. The youngest and oldest sample member strata are relatively more homogeneous with respect to employment status.

For the detection of a risk of lower response quality, only the validation data and paradata approaches are options. The LFS questionnaire has a complex routing and is not structured around latent variables or constructs. Since rich administrative data are available, we consider the validation data option. Auxiliary variables can be linked from government employment and unemployment administrative data. Three comparisons are made

- Difference 1: Not employed in register, but employed in LFS
- Difference 2: Not employed in register and no employment office registration, but subscription to employment office in LFS
- Difference 3: Employed in register, but not employed in LFS

Around 8.0% of the respondents shows one of the three differences, when including both self-reporting and proxy reporting. The first and second differences are largely unrelated to the type of reporting. However, the third difference increases from 3.7% to 4.6%, when proxy reporting is allowed. Despite this relatively modest increase of 0.9%, we explore a response quality indicator for a type 3 difference. Table 11.6 contains the propensities for the three age groups for difference 3. Table 11.6 shows that the largest differences are found for the youngest group; the proportion of respondents that deviate from administrative data increases from 6.0% to 7.8%. Therefore, it appears that proxy reporting for this group is more prone to error than for other age groups.

We explore an optimization problem in which the response rate is optimized against constraints on the *R*-indicator given age, the budget, and the response quality rate. The budget is measured in terms of the expected number of face-to-face visits needed. The optimization problem is formulated as

$$\max_{p_g} \rho \quad \text{subject to the constraints}$$

$$R \geq \alpha, \quad C \leq B, \quad rq \leq \Gamma,$$

TABLE 11.6

Estimated Response Quality Propensities Per Age Group and Overall for Type 3 Differences

Design Feature	15–25 years (%)	26–55 years (%)	56–65 years (%)	All (%)
Self-reporting only	6.0	2.8	4.1	3.7
Proxy allowed	7.8	3.5	4.7	4.6

TABLE 11.7

Optimal Response Rates for Different Levels of the Cost Constraint and the *R*-Indicator, When the Response Quality Constraint Is Set at 3.5%

Budget Level	*R*-Indicator	Optimal Response Rate
3.0 visits	0.80	60.9%
2.5 visits	0.80	49.3%
3.0 visits	0.85	60.9%
2.5 visits	0.85	49.3%
3.0 visits	0.90	Not feasible
2.5 visits	0.90	Not feasible

Note: *R*-indicator constraints of 0.90 lead to infeasible optimization problems.

$$\pi_g \in [0,1],\ p_g(a) \in [0,1], \text{ and } \sum_{a=1}^{A} p_g(a) = 1,$$

where $\alpha \in \{0.80, 0.85, 0.90\}$, $\Gamma = 3.5\%$, and budget is $B \in \{2.5, 3.0\}$. Hence, a maximum of 3.5% differences to administrative data of type 3 is allowed while not exceeding either 2.5 or 3.0 visits on average and constraining the variance in age response propensities.

Table 11.7 shows the maximal response rates given the two levels of the budget and the three *R*-indicator constraints. From Table 11.6 we can see that the response quality rate constraint of 3.5% is very stringent; only the middle-age stratum has a response quality propensity that stays below this level under proxy reporting. The other two age groups have even higher values than 3.5% for self-reporting, and, hence, need to be assigned mostly to self-reporting in order not to overrun the overall 3.5%. As a consequence, in the optimization, the middle-age stratum gets overrepresented in order to maximize the response rate, but at the cost of overrepresentation. For an *R*-indicator value of 0.90 there are simply no solutions that keep the response quality rate below 3.5%. An *R*-indicator of 0.85 can be reached and further lowering the *R*-indicator to 0.80 has no impact; the response quality rate constraint has become the dominant constraint. The only option to improve representativeness is to raise the response quality rate to values above 6.0%; for this level, all age strata have a self-reporting response quality propensity below the threshold. The more restrictive constraint on 2.5 visits per address leads to a considerable drop in the response rate, but not in the allocation to self-reporting and proxy reporting.

11.4 Summary

In this chapter, we extended ASD framework to include measurement error for single-purpose surveys and for multi-purpose surveys. The extended framework is a first start and more research is needed. We end with a number of observations:

- Measurement error introduces a very different perspective to ASD. Measurement error is inherently hard to estimate without repeated measurements or experimental designs, and cannot be estimated or evaluated solely on the basis of related auxiliary data. In extending ASDs, a distinction needs to be made between single-purpose surveys and multi-purpose surveys.
- ASDs for single-purpose surveys may focus directly on the combined effect of nonresponse and measurement error against a specified benchmark design. Such benchmark designs may correspond to counterfactual designs in which one strategy is deemed optimal in terms of nonresponse, while another strategy is deemed best in measurement.
- ASDs for multi-purpose surveys may focus on indirect indicators of response quality. Response quality propensities may be estimated from linked validation data, paradata, and/or latent variable models.
- Unlike nonresponse, measurement error does not lend itself for adaptation during data collection. It is possible but not at all customary to contact respondents when a risk of measurement error is detected. However, in a panel or cohort setting response quality in previous waves may be used to inform decisions.

12

The Future of Adaptive Survey Design

ASD springs out of a particular context. That context includes several challenges for surveys, such as declining response, rising costs, and an emergence of the Internet as a new mode of interviewing. ASD can be seen as a response to these challenges. Of course, there are precedents for adaptive designs. Some of these were discussed in Chapter 1. But the challenges of the current situation have called for a more thoroughgoing rethinking of current procedures. ASD provides a unifying framework for this reevaluation.

Of course, this rethinking is still in progress and the context in which surveys operate will continue to evolve. Therefore, we should expect that ASD will continue to evolve. In this chapter, we describe some of the key challenges to ASD and areas that we anticipate will be the focus of further investigation and development. We find nine areas where further changes, methodological development, and experience are needed. The first area relates to the utility of ASD. The second, third, fourth, and fifth relate to the implementation of ASD. The sixth, seventh, eighth, and ninth refer to ASD methodology.

1. *Evidence for what type of ASDs have been found effective and in what settings*: While evidence has been gradually accumulating, there is great need for more empirical tests of ASDs, as both survey settings (e.g., population, duration, mode, and topic) and objectives (e.g., nonresponse rates, nonresponse bias, measurement error, and cost minimization) create substantial variation in ASDs and their outcomes. The empirical evidence is still limited in terms of number of studies and types of designs.

   To date, studies have demonstrated that response propensities can be estimated well, interventions can be devised that are effective at increasing participation among targeted groups, and survey representativeness can be increased. Most designs, however, rely on changing protocols and, in particular, escalating the protocols for targeted cases or stopping data collection effort for nontargeted cases. Yet, rather than the implicit escalation of protocols (e.g., Web to mail to telephone to in-person), protocols could be seen as alternatives, as in the work by Schouten, Calinescu, and Luiten (2013) described in Chapter 11. Subgroups can respond differently to different design features. Some respondents may provide less measurement error in self-administered modes while other respondents may do so in interviewer-administered modes. Some may respond to a Web survey, others may not. ASDs offer the framework for tailoring the protocol to different sample members, and there is the need for studies to demonstrate such tailoring in surveys.

ASDs have been criticized for balancing survey response on auxiliary data that can be used in a cheaper and equally effective manner in the postsurvey adjustment. There is recent literature (Särndal, Lumiste, and Traat, 2016; Schouten et al., 2016) that discusses this criticism and conditions under which corrections through survey design are more effective than corrections in estimation. However, more research is clearly needed.

Dissemination of methods and results should also be accessible by a diverse audience. Designs and results should be described in a manner that can be understood by individuals with different roles in the survey, from project managers to people in data collection operations. Similarly, it may also be beneficial to use multiple channels for sharing of information—for example, not limited to scientific journals. Broader accessibility of the results should facilitate future implementation of ASDs, and could improve the designs and adherence to protocols.

2. *Practical approaches, procedures, and statistical methods that facilitate implementation of ASDs*: ASDs need to be tailored to each survey, but there are procedures that could be shared. These procedures and approaches are different than the skills described in (5) below, and the technical systems described in (4). Here, we mean that the toolkit of methods, approaches, and previous examples needs to be expanded. For example, we discussed the challenge in deciding to use data from a prior implementation of the survey or from the current implementation, to inform which cases should start receiving another treatment. Yet, Bayesian methods could be used to incorporate both sources of information, and allow the data from the current implementation to gradually become more influential over the course of data collection. Making such procedures available to the research community could facilitate the implementation of increasingly complex ASDs.

   Approaches to monitoring and estimating costs may be another area where new methods can be developed. Apart from the creation of technical systems that track costs directly, new methods for estimating cost parameters for adaptive designs—often at a highly detailed level—are needed. This may include statistical estimation of parameters that are not directly observed. It may also consider complicated problems, such as marginal costs for each unit in cluster samples.

Tools for monitoring are another area where collaboration within and across survey organizations can lead to improved procedures. In part, this is a technical problem that involves data management, software and systems for the development of monitoring tools, such as "dashboards." But there are additional problems, including methods for presenting results, especially incorporating measures of uncertainty into these displays. There is also a need to develop indicators for other potential sources of error. We have discussed indicators for nonresponse in Chapter 9, and indicators for measurement error—albeit in a more limited fashion—in Chapter 11. These indicators could be extended, but indicators for coverage error, sampling error, and even postprocessing errors would be useful. Development of these error indicators, along with cost indicators, will improve our ability to implement optimization of ASD under a total survey error perspective.

3. *Definition and flexibility of survey objectives and constraints*: The survey objectives and constraints can be inconsistent, which limits the potential impact of ASD. Surveys often try to increase the precision of survey estimates through sample stratification, targeting of subgroups, and weight trimming. Yet, surveys tend to have constraints expressed not only in terms of cost, but also number of interviews. A design that minimizes variances under cost constraints is unlikely to be the same as a design that does so with an additional requirement of a number of completed interviews. Some interviews may disproportionately contribute to the precision of an estimate, but may also cost more than other interviews. As a result, greater precision may be achieved for the same overall cost, *but with fewer interviews*. This suggests that survey sponsors should specify the requirements for the survey using precision (e.g., standard error) rather than number of completed interviews.

   It may help to consider current trends. Chapter 2 described increasing costs of survey data collection. Consider a survey that is required to continue to produce the same number of interviews every year. If the survey budget remains constant, the survey may have to collect interviews through less intensive methods, or collect more interviews from strata associated with lower data collection costs. While the requirement for a fixed number of interviews may simplify contracts between a funding organization and a data collection organization, relaxing this constraint could lead to ASDs that better achieve the survey objectives.

   Number of interviews is only an important example—there are other study-specific constraints, such as number of interviews by phase and sample allocation by frame, that limit what can be achieved with an ASD. Specifying the key objectives, as described in Chapter 2, should help. The survey can then be designed with these key objectives in mind.

Just as in (1) above—building evidence for the effectiveness of ASD—some of the burden falls on those involved in ASD to disseminate results that are widely accessible. If the individuals responsible for setting constraints on a survey become more familiar with ASDs, then the imposed constraints may allow for adaptive designs that can have greater impact.

4. *Versatile data collection systems*: An ASD operates within the constraints imposed by software, such as sample management systems. These systems have been rapidly evolving in recent times. For example, a mixed-mode ASD may require that sample cases can be moved from one mode to another based on rules. Furthermore, sample cases may have to be accessible from multiple modes regardless of current mode assignment. Sample management systems also need to be able to accommodate multiple rules. These rules can take different forms; examples of complex rules include the following:
   a. If a sample member refuses on the phone, then the case may be put "on hold" for a set number of days before a refusal conversion call may be attempted.
   b. If the sample member answers in Spanish, then the case would also have to be assigned to a Spanish-speaking interviewer, but if the number of call attempts resulting in noncontact exceeds some limit, then the case may be assigned to another mode.

   Both the complexity of these systems and our ability to set sophisticated sets of rules will likely increase over time. Both the systems and the rules will evolve over time—as the systems expand capabilities, new rules will be tried, further extensions of the systems may be proposed, leading to the elaboration of potential new rules, and so on. Given these evolving requirements, flexibility will be a key attribute of new technical systems for the management of data collection.

5. *An extended skillset to fulfill the new needs for survey design, data collection operations, and new roles on the survey*: Starting with Chapter 2, we have described how ASD can be different for each survey based on the study objectives and its design features, such as instrument, mode, target population, sampling frame, etc. Chapters 4 and 7 noted that we usually do not know what the "optimum" ASD may be—we need to design and test different interventions and further optimize the design. Designing and testing alternative treatments at the sample unit level has not been a usual part of the survey practitioner's task in the past. Thus, investigators and study directors

will need to consider what skills will be needed. From the designs that have been described so far, it is apparent that a combination of a strong statistical background and knowledge of data collection methods is essential. These skills may not necessarily be contained in a single person, but distributed across staff members. However, these staff members would need to work together in a process that includes all the necessary skills. This may lead to a new cadre of staff or additional training for current staff who could incorporate ASD plans prior to survey implementation.

Apart from the skillset, the role of the survey statistician may have to change. In a traditional static survey design, the key roles of the statistician are prior to data collection, in sample design and selection, and after data collection, in weighting and estimation. Some limited monitoring of data collection progress may also occur, for example, to determine whether additional sample should be released. In studies with ASD, the survey statistician's role is generally far more involved. This includes helping design strata for the different interventions as described in Chapter 3, and for a heavier role in monitoring during data collection, as described in Chapter 5. The design may call for continuous monitoring of paradata and survey measures, and for estimation of models to identify sample cases for intervention.

In addition to the different skills and involvement of survey statisticians, ASDs can also demand greater collaboration across departments within the survey organization. For example, the statistician can determine which sample members need a different intervention, but not know what that intervention needs to be. Similarly, those involved in data collection need to understand the design and goals of an ASD, in order to facilitate and potentially improve its implementation. Therefore, close collaboration is needed, starting from the design stage of the survey. This collaboration may require new management structures and new ways of interacting during the design and implementation phases as described in Chapter 6 on logistics.

For some interventions, the ASD can be counter to common practice, such as telling interviewers which sample cases to pursue in a given week—when common practice is for interviewers to decide which sample cases to attempt and on what days. Additional training and greater staff commitment would be needed in order to have interviewers comply with the ASD protocol.

6. *Designed paradata*: Paradata have a critical role in ASD. To use one of the examples in this book, if we are deciding which cases to stop as being unproductive, we would need data on the outcomes from previous contact attempts, the cost per contact attempt if it varies in

a particular study, and any additional auxiliary data available on the sample. While more data provide for better informed designs, there are at least three major challenges to obtaining and using paradata. First, some paradata are more difficult to collect. For example, even a seemingly straight-forward measure such as the outcome of a contact attempt can be problematic. Some contact attempts may not even be recorded (e.g., Biemer, Chen, and Wang, 2013). In those cases where a refusal occurs, the reasons for a refusal recorded by the interviewer are also subject to interviewer error. Such errors, if random, diminish the utility of the collected paradata; if there are systematic errors, such as interviewers in one geographic area being more likely to not record a contact attempt or misreport the outcome, these errors can lead to misallocation of resources in an ASD. ASDs rely on the quality of the auxiliary data, and greater attention to how paradata are collected is needed.

Second, there is the need for designed paradata; the data that are being collected may not be sufficient. For example, in a survey with limited substantive information on the sampling frame, paradata may need to be designed for that particular survey in order to intervene on sample households in a manner that could reduce nonresponse bias. An example of this is the NSFG, which asks interviewers to rate whether the household informant is sexually active, as this observation is related to key variables in the survey (West, 2013). Another example is the California Health Interview Survey, which included two questions in the screener, one on health conditions and another on health insurance, with the sole purpose of informing about nonresponse bias during data collection (Peytcheva, Peytchev, and Jans, 2016). Although very few surveys have attempted it, this practice shows substantial promise to inform ASDs.

Third, the paradata may need to be modified in order to serve a particular purpose. For instance, cost may be needed at the sample unit level in an in-person survey. Such data are not readily available, as interviewers have multiple sample households and generally could not assign time spent on each sample household. The marginal cost of a call attempt at a household would be needed for an adaptive design that incorporates cost differences across sample households. Similarly, some sample telephone numbers in a telephone survey may receive a different treatment and it may be essential to evaluate the cost effectiveness of the treatment. Yet, the interviewing cost may be collected only at the study level. In order to measure cost differences, either the sample would need to be separated into separate projects, or calculations would need to be made that estimate cost from the number of call attempts and types of outcomes.

7. *Longitudinal survey designs*: We illustrated many of the ideas using examples from repeated cross-sectional surveys but longitudinal survey designs have especially desirable features for ASD as they amass data across multiple points in time. Models such as for optimum mode of initial contact and optimum mode for collecting data can be informed by data from previous iterations (waves). Furthermore, each wave of data collection can be viewed as an opportunity to learn about which protocol may be more desirable for a particular respondent, incorporating measures that are intended to inform the tailoring in the following wave—for example, asking respondents whether they were reluctant due to time constraints, lack of interest, or something else, in order to determine whether to assign a shorter instrument, a higher incentive amount, or some other treatment in a later data collection. Of particular interest may be studies that implement more intensive measurement (e.g., daily). These studies present a real opportunity to learn from the behavior of individual respondents and tailor the protocol over repeated attempts to measure them.
8. *Multiple sources of error*: ASDs have so far focused predominantly on reducing nonresponse and cost. This is not a limitation of ASDs, but merely reflects the greatest concerns that these studies have encountered. It would be unfortunate if ASD is seen only as a way to reduce cost. ASDs can be used to address multiple sources of error; as we described in Chapter 11. For example, nonresponse and measurement error could be considered simultaneously in the design. Eventually, other error sources could be included. We can imagine adaptive designs aimed at balancing multiple sources of error, including not only nonresponse error, measurement error, and sampling error, but coverage error, postprocessing error, and others.
9. *Optimization of ASD*: ASD, more than uniform survey design, leans heavily on optimization, that is, on making trade-offs between objectives and constraints. Given that designers differentiate effort between different sample units, it is imperative that the effectiveness and costs of treatments can be estimated accurately. Optimization problems are, however, often nonlinear and nonconvex, so that numerical methods are needed for which there is no guaranteed convergence or convergence to a global optimum. Clever optimization routines that build in knowledge about likely optimal solutions are key. Research in this area is still sparse and may learn from other statistical areas like dynamic treatment regimes.

These are only a few areas of ASD that we think should receive greater attention. The proliferation of paradata and other auxiliary information, increased ability to harness these data, more sophisticated real-time data collection systems, and developments in statistical methods—all in the context of rising challenges to surveys—open great opportunities for ASDs. We aimed to provide an understanding of ASD, to elaborate on each major component, and to provide some essential technical background. The possibilities, however, are not limited to what we have presented, and we look forward to the future developments in ASD.

References

Adams, S. A., C. E. Matthews, C. B. Ebbeling, C. G. Moore, J. E. Cunningham, J. Fulton, and J. R. Hebert. The effect of social desirability and social approval on self-reports of physical activity. *American Journal of Epidemiology* 161(4); 2005: 389–98.

Agresti, A. *Categorical Data Analysis* (2nd ed.). New York: John Wiley & Sons, 2002.

Alwin, D. F. *Margins of Error.* Hoboken, NJ: Wiley, 2007.

Andresen, E. M., C. R. Machuga, M. E. Van Booven, J. Egel, J. T. Chibnall, and R. C. Tait. Effects and costs of tracing strategies on nonresponse bias in a survey of workers with low-back injury. *Public Opinion Quarterly* 72(1); 2008: 40–54.

Andridge, R. R. and R. J. A. Little. Proxy pattern-mixture analysis for survey nonresponse. *Journal of Official Statistics* 27(2); 2011: 153–80.

Bais, F., B. Schouten, P. Lugtig, V. Toepoel, J. Arends-Toth, S. Douhou, N. Kieruj et al. Can Survey Item Characteristics Relevant to Mode-Specific Measurement Error Be Coded Reliably? A Case Study on Eleven Dutch General Population Surveys, under review with Sociological Methods and Research, in press.

Bang, H. and J. M. Robins. Doubly robust estimation in missing data and causal inference models. *Biometrics* 61; 2005: 962–72.

Bather, J. *Decision Theory: An Introduction to Dynamic Programming and Sequential Decisions.* New York, NY: John Wiley, 2000.

Baumgartner, H. and J. E. M. Steenkamp. Response styles in marketing research: A cross-national investigation. *Journal of Marketing Research* 28; 2001: 143–56.

Beaumont, J.-F., D. Haziza and C. Bocci. An adaptive data collection procedure for call prioritization. *Journal of Official Statistics* 30; 2014: 607–21.

Beckett, M. K., M. N. Elliott, S. Gaillot, A. Haas, J. W. Dembosky, L. A. Giordano, and J. Brown. Establishing limits for supplemental items on a standardized national survey. *Public Opinion Quarterly* 80(4): 964–76.

Beebe, T. J., G. Richard Locke III, S. A. Barnes, M. E. Davern, and K. J. Anderson. Mixing web and mail methods in a survey of physicians. *Health Services Research* 42(3p1); 2007: 1219–34.

Belsley, D. A., E. Kuh, and R. E. Welsch. *Regression Diagnostics: Identifying Influential Observations and Sources of Collinearity,* Vol. 101. New York, NY: John Wiley & Sons, 1980.

Berry, D. A. Adaptive trials and bayesian statistics in drug development. *Biopharmaceutical Report* 9(2); 2001: 1–7.

Bertsekas, D. P. *Non-Linear Programming.* Belmont, MA: Athena Scientific, 1995.

Bertsima, D. P. and J. N. Tsitsiklis. *Introduction to Linear Optimization.* Belmont, MA: Athena Scientific, 1997.

Bethlehem, J. Reduction of nonresponse bias through regression estimation. *Journal of Official Statistics* 4; 1988: 251–60.

Bethlehem, J. G., F. Cobben, and J. G. Schouten: *Handbook of Nonresponse in Household Surveys, Handbook Wiley Series in Survey Methodology.* New York: Wiley, (p. 467). 2011.

Biemer, P. P. *Latent Class Analysis of Survey Error.* New York, NY: John Wiley & Sons, 2010.

Biemer, P. P., P. Chen, and K. Wang. Using level-of-effort paradata in non-response adjustments with application to field surveys. *Journal of the Royal Statistical Society: Series A (Statistics in Society)* 176(1); 2013: 147–68.

Biemer, P. P. and L. Stokes. Approaches to modeling of measurement errors, In *Measurement Error in Surveys*, edited by P. P. Biemer, R. M. Groves, L. E. Lyberg, N. A. Mathiowetz, and S. Sudman (pp. 487–517). New York: Wiley, 1991.

Bollapragada, S. and S. K. Nair. Improving right party contact rates at outbound call centers. *Production and Operations Management* 19(6); 2010: 769–79.

Borkan, B. The mode effect in mixed-mode surveys: Mail and web surveys. *Social Science Computer Review* 28(3); 2010: 371–80.

Brick, J. M. Unit nonresponse and weighting adjustments: A critical review. *Journal of Official Statistics* 29(3); 2013: 329–53.

Brick, J. M., B. Allen, P. Cunningham, and D. Maklan. Outcomes of a calling protocol in a telephone survey. *Proceedings of the Survey Research Methods Section of the American Statistical Association* 1996: 142–9.

Brick, J. M., W. R. Andrews, P. D. Brick, H. King, N. A. Mathiowetz, and L. Stokes. Methods for improving response rates in two-phase mail surveys. *Survey Practice* 5(3); 2012.

Brick, J. M., I. F. Cervantes, S. Lee, and G. Norman. Nonsampling errors in dual frame telephone surveys. *Survey Methodology* 37(1); 2011a: 1–12.

Brick, J. M. and M. E. Jones. Propensity to respond and nonresponse bias. *METRON—International Journal of Statistics* LXVI(1); 2008: 51–73.

Brick, J. M., D. Judkins, J. Montaquila, and D. Morganstein. Two-phase list-assisted RDD sampling. *Journal of Official Statistics* 18(2); 2002: 203–15.

Brick, J. M., D. Williams, and J. M. Montaquila. Address-based sampling for subpopulation surveys. *Public Opinion Quarterly* 75(3); 2011b: 409–28.

Bruin, L., B. Schouten, N. Mushkudiani, N. Shlomo, S. Coffey, G. Durrant, P. Lundquist et al. A Bayesian analysis of design parameters in survey data collection. CBS Discussion Paper, The Hague, The Netherlands, 2016. Available at www.cbs.nl

Burger, J., K. Perryck, and B. Schouten. Robustness of adaptive survey designs to inaccuracy of design parameters. *Journal of Official Statistics*, in press.

Calinescu, I. M. Optimal resource allocation in adaptive survey designs, PhD thesis, Vrije Universiteit, 2013.

Calinescu, M., S. Bhulai, and B. Schouten. Optimal resource allocation in survey designs. *European Journal of Operational Research* 226(1); 2013: 115–21.

Calinescu, M. and B. Schouten. Adaptive survey designs to minimize mode effects. A case study on the Dutch Labour Force Survey. *Survey Methodology* 41(2); 2015: 403–25.

Calinescu, M. and B. Schouten. Adaptive survey designs for nonresponse and measurement error in multi-purpose surveys. *Survey Research Methods* 10(1); 2016: 35–47.

Campanelli, P., P. Sturgis, and S. Purdon. *Can You Hear Me Knocking?: Investigation into the Impact of Interviewers on Survey Response Rates*. London: National Centre for Social Research, 1997.

Chapman, C. National Center for Education Statistics Adaptive Design Overview. Paper presented at the Federal Committee on Statistical Methodology Conference, Washington, DC, 2014.

Chesnut, J. Model-based mode of data collection switching from internet to mail in the American Community Survey. *2013 American Community Survey Research and Evaluation Report Memorandum Series #ACS13-RER-18*, Boston, MA. Washington, DC: US Census Bureau, 2013, pp. 1–17.

Cochran, W. G. *Sampling Techniques* (3rd ed.). New York, NY: Wiley, 1977.

Collins, L. M., S. A. Murphy, and V. Strecher. The multiphase optimization strategy (Most) and the sequential multiple assignment randomized trial (Smart): New methods for more potent ehealth interventions. *American Journal of Preventive Medicine* 32(5 Supplement 1); 2007: S112–18.

Cominole, M., A. Peytchev, D. J. Pratt, B. Shepherd, P. Siegel, D. Wilson, and J. Wine. Using Mahalanobis distance measures for bias reduction. American Association for Public Opinion Research Annual Conference, Boston, MA, 2013.

Couper, M. P. Measuring survey quality in a CASIC environment. *Proceedings of the Survey Research Methods Section of the American Statistical Association*, Dallas, TX, 1998, pp. 41–49.

Couper, M. P. *Designing Effective Web Surveys*. Cambridge, New York: Cambridge University Press, 2008.

Couper, M. and M. B. Ofstedal. Keeping in contact with mobile sample members. In *Methodology of Longitudinal Surveys*, edited by Peter Lynn (pp. 183–203). Chichester, UK: Wiley, 2009.

Couper, M. P. and J. Wagner. Using paradata and responsive design to manage survey nonresponse. *World Statistics Congress of the International Statistical Institute Conference*, Dublin, Ireland, 2011, pp. 542–548.

Curtin, R., S. Presser, and E. Singer. The effects of response rate changes on the index of consumer sentiment. *Public Opinion Quarterly* 64(4); 2005: 413–28.

Curtin, R., S. Presser, and E. Singer. Changes in telephone survey nonresponse over the past quarter century. *Public Opinion Quarterly* 69(1); 2000b: 87–98.

De Heij, V., B. Schouten, and N. Shlomo. RISQ 2.1 Manual. Tools in SAS and R for the Computation of R-Indicators and Partial R-Indicators, 2015. Available at www.risq-project.eu

de Leeuw, E. D. To mix or not to mix data collection modes in surveys. *Journal of Official Statistics* 21(2); 2005: 233–255.

Deutskens, E., K. de Ruyter, M. Wetzels, and P. Oosterveld. Response rate and response quality of internet-based surveys: An experimental study. *Marketing Letters* 15(1); 2004: 21–36.

Dillman, D. A., G. Phelps, R. Tortora, K. Swift, J. Kohrell, J. Berck, and B. L. Messer. Response rate and measurement differences in mixed-mode surveys using mail, telephone, interactive voice response (IVR) and the internet. *Social Science Research* 38(1); 2009: 1–18.

Dillman, D. A., M. D. Sinclair, and J. R. Clark. Effects of questionnaire length, respondent-friendly design, and a difficult question on response rates for occupant-addressed census mail surveys. *Public Opinion Quarterly* 57(3); 1983: 289–304.

Dillman, D. A., M. D. Sinclair, and J. R. Clark. Effects of questionnaire length, respondent-friendly design, and a difficult question on response rates for occupant-addressed census mail surveys. *Public Opinion Quarterly* 57(3); 1993: 289–304.

Dillman, D. A., J. D. Smyth, and L. M. Christian. *Internet, Mail, and Mixed-Mode Surveys: The Tailored Design Method* (4th ed.). Hoboken, NJ: Wiley & Sons, 2014.

Dotinga, A., R. J. J. M. Van den Eijnden, W. Bosveld, and H. F. L. Garretsen. The effect of data collection mode and ethnicity of interviewer on response rates and self-reported alcohol use among turks and moroccans in the Netherlands: An experimental study. *Alcohol and Alcoholism* 40(3); 2005: 242–8.

Durrant, G. B. and F. Steele. Multilevel modelling of refusal and non-contact in household surveys: Evidence from six UK Government surveys. *Journal of the Royal Statistical Society: Series A (Statistics in Society)* 172(2); 2009: 361–81.

Dykema, J., K. Jaques, K. Cyffka, N. Assad, R. G. Hammers, K. Elver, K. C. Malecki, and J. Stevenson. Effects of sequential prepaid incentives and envelope messaging in mail surveys. *Public Opinion Quarterly* 79(4); 2015: 906–931.

Dykema, J., J. Stevenson, L. Klein, Y. Kim, and B. Day. Effects of e-mailed versus mailed invitations and incentives on response rates, data quality, and costs in a web survey of university faculty. *Social Science Computer Review* 31(3); 2015: 359–70.

Earp, M. and J. McCarthy. Using nonresponse propensity scores to improve data collection methods and reduce nonresponse bias. *Paper presented at the AAPOR*, Phoenix, AZ, 2011.

Edwards, P., I. Roberts, M. Clarke, C. DiGuiseppi, S. Pratap, R. Wentz, and I. Kwan. Increasing response rates to postal questionnaires: Systematic review. *British Medical Journal* 324(7347); 2002: 1183.

Few, S. *Information Dashboard Design: The Effective Visual Communication of Data* (1st ed.). Sebastopol, CA: O'Reilly, 2006.

Fumagalli, L., H. Laurie, and P. Lynn. Experiments with methods to reduce attrition in longitudinal surveys. *Journal of the Royal Statistical Society: Series A (Statistics in Society)* 176(2); 2013: 499–519.

Galili, T. Modeling and Solving Linear Programming with R, 2015. Available at www.omniascience.com, USA.

GAO. *2020 Census: Progress Report on the Census Bureau's Efforts to Contain Enumeration Costs*. Washington, DC: United States Government Accountability Office, 2013.

Gelman, A., J. B. Carlin, H. S. Stern, and D. B. Rubin. *Bayesian Data Analysis* (2nd ed.). London: Chapman & Hall, 2014.

Gelman, A. and J. Hill. *Data Analysis Using Regression and Multilevel/Hieracrhical Models*. UK: Cambridge University Press, 2007.

Gfroerer, J., Eyerman, J., and Chromy, J. (Eds.). *Redesigning an Ongoing National Household Survey: Methodological Issues*. DHHS Publication No. SMA 03-3768. Rockville, MD: Substance Abuse and Mental Health Services Administration, Office of Applied Studies, 2002.

Greenberg, B. S. and S. L. Stokes. Developing an optimal call scheduling strategy for a telephone survey. *Journal of Official Statistics* 6(4); 1990: 421–35.

Groves, R. M. *Survey Errors and Survey Costs, Wiley Series in Probability and Mathematical Statistics. Applied Probability and Statistics*. New York: Wiley, 1989.

Groves, R. M., J. M. Brick, M. P. Couper, W. Kalsbeek, B. Harris-Kojetin, F. Kreuter, B.-E. Pennell et al. Issues facing the field: Alternative practical measures of representativeness of survey respondent pools. *Survey Practice* 1(3) (October): 2008. Accessed March 10, 2009. http://surveypractice.org/index.php/SurveyPractice/article/view/221/html.

Groves, R. M., M. P. Couper, S. Presser, E. Singer, R. Tourangeau, G. P. Acosta, and L. Nelson. Experiments in producing nonresponse bias. *Public Opinion Quarterly* 70(5); 2006: 720–36.

Groves, R. M. and S. Heeringa. Responsive design for household surveys: Tools for actively controlling survey errors and costs. *Journal of the Royal Statistical Society Series A: Statistics in Society* 169(Part 3); 2006: 439–57.

Groves, R. and K. A. McGonagle. A theory-guided interviewer training protocol regarding survey participation. *Journal of Official Statistics* 17(2); 2001: 249–66.

Groves, R. M. and E. Peytcheva. The impact of nonresponse rates on nonresponse bias: A meta-analysis. *Public Opinion Quarterly* 72(2); 2008: 167–89. doi: 10.1093/poq/nfn011.

Groves, R. M., S. Presser, and S. Dipko. The role of topic interest in survey participation decisions. *Public Opinion Quarterly* 68(1); 2004: 2–31.

Groves, R. M., E. Singer, and A. Corning. Leverage-saliency theory of survey participation—Description and an illustration. *Public Opinion Quarterly* 64(3); 2000: 299–308.

Hansen, M. H. and W. N. Hurwitz. The problem of non-response in sample surveys. *Journal of the American Statistical Association* 41(236); 1946: 517–29.

Hansen, M. H., W. N. Hurwitz, and W. G. Madow. *Sample Survey Methods and Theory.* New York, NY: Wiley, 1953.

Hartley, H. O. *Multiple Frame Surveys.* American Statistical Association, Section on Social Statistics. Washington, DC, 1962, pp. 203–206.

Hastie, T., R. Tibshirani, and J. Friedman. *The Elements of Statistical Learning, Second Edition: Data Mining, Inference, and Prediction* (2nd ed.). New York, NY: Springer, 2009.

Heerwegh, D. Explaining response latencies and changing answers using client-side paradata from a web survey. *Social Science Computer Review* 21(3); 2003: 360–73.

Heerwegh, D. Mode differences between face-to-face and web surveys: An experimental investigation of data quality and social desirability effects. *International Journal of Public Opinion Research* 21(1); 2009: 111–21.

Hirano, K., G. Imbens, and G. Ridder. Efficient estimation of average treatment effects using the estimated propensity score. *Econometrica* 71(4); 2003: 1161–89.

Horvitz, D. and D. Thompson. A generalization of sampling without replacement from a finite population. *Journal of the American Statistical Association* 47; 1952: 663–85.

Hox, J. J., E. D. de Leeuw, and H.-T. Chang. Nonresponse versus measurement error: Are reluctant respondents worth pursuing? *Bulletin of Sociological Methodology/Bulletin de Mèthodologie Sociologique* 113(1); 2012: 5–19.

Johnson, T. P., M. Fendrich, C. Shaligram, A. Garcy, and S. Gillespie. An evaluation of the effects of interviewer characteristics in an RDD telephone survey of drug use. *Journal of Drug Issues* 30(1); 2000: 77–101.

Kalsbeek, W. D., S. L. Botman, J. T. Massey, and P.-W. Liu. Cost-efficiency and the number of allowable call attempts in the national health interview survey. *Journal of Official Statistics* 10(2); 1994: 133–52.

Kalsbeek, W. D., R. E. Folsom, Jr., and A. F. Clemmer: The national assessment no-show study: An examination of nonresponse bias. *Proceedings of the Social Statistics Section,* St. Louis, MO. Washington, DC: American Statistical Association, 1974, pp. 180–189.

Kalton, G. and I. Flores-Cervantes. Weighting methods. *Journal of Official Statistics* 19(2); 2003: 81–97.

Kaplowitz, M. D., F. Lupi, M. P. Couper, and L. Thorp. The effect of invitation design on web survey response rates. *Social Science Computer Review* 30(3); 2012: 339–49.

Karp, J. A. and D. Brockington. Social desirability and response validity: A comparative analysis of overreporting voter turnout in five countries. *Journal of Politics* 67(3); 2005: 825–40.

Keeter, S., C. Miller, A. Kohut, R. M. Groves, and S. Presser. Consequences of reducing nonresponse in a national telephone survey. *Public Opinion Quarterly* 64; 2000: 125–48.

Kirgis, N. and J. Lepkowski. Design and management strategies for paradata-driven responsive design: Illustrations from the 2006–2010 national survey of family growth. In *Improving Surveys with Paradata: Analytic Uses of Process Information*, edited by F. Kreuter. Hoboken, NJ: Wiley, 2013, pp. 121–144.

Kish, L. *Survey Sampling*. New York, NY: Wiley, 1965.

Klausch, L. T., J. J. Hox, and B. Schouten. Measurement effects of survey mode on the equivalence of attitudinal rating scale questions. *Sociological Methods & Research* 42(3); 2013: 227–63.

Klofstad, C. A., S. Boulianne, and D. Basson. Matching the message to the medium: Results from an experiment on internet survey email contacts. *Social Science Computer Review* 26(4); 2008: 498–509.

Kreuter, F. *Improving Surveys with Paradata: Analytic Uses of Process Information*, edited by F. Kreuter. Hoboken, NJ: John Wiley & Sons, 2013.

Kreuter, F., A. Mercer, and W. Hicks. Increasing fieldwork efficiency through prespecified appointments. *Journal of Survey Statistics and Methodology* 2(2); 2014: 210–23.

Kreuter, F., S. Presser, and R. Tourangeau. Social desirability bias in Cati, Ivr, and web surveys the effects of mode and question sensitivity. *Public Opinion Quarterly* 72(5); 2008: 847–65.

Kristal, A. R., E. White, J. R. Davis, G. Corycell, T. Raghunathan, S. Kinne, and T.-K. Lin. Effects of enhanced calling efforts on response rates, estimates of health behavior, and costs in a telephone health survey using random-digit dialing. *Public Health Reports* 108(3); 1993: 372.

Laflamme, F. and M. Karaganis. Development and implementation of responsive design for CATI Surveys at Statistics Canada. Paper presented at the European Conference on Quality in Official Statistics (Q2010), Helsinki, Finland, 2010.

Lepkowski, J. M. and M. P. Couper. Nonresponse in the second wave of longitudinal household surveys. In *Survey Nonresponse*, edited by R. Groves, D. Dillman, J. Eltinge, and R. J. A. Little (pp. 259–72). New York: Wiley, 2002.

Lepkowski, J. M., W. D. Mosher, R. M. Groves, B. T. West, J. Wagner, and H. Gu. Responsive design, weighting, and variance estimation in the 2006–2010 National Survey of Family Growth. In *Vital and Health Statistics, 2*. National Center for Health Statistics. Hyattsville, MD, 2013.

Little, R. J. A. and D. B. Rubin. *Statistical Analysis with Missing Data, Wiley Series in Probability and Statistics*. Wiley: New York, 2002.

Lipps, O. A note on improving contact times in panel surveys. *Field Methods* 24(1); 2012: 95–111.

Luiten, A. and B. Schouten. Tailored fieldwork design to increase representative household survey response: An experiment in the survey of consumer satisfaction. *Journal of the Royal Statistical Society: Series A (Statistics in Society)* 176(1); 2013: 169–89.

Lundquist, P. and C. E. Särndal. Aspects of responsive design for the Swedish Living Conditions Survey. *Journal of Official Statistics* 29; 2013: 557–82.

Lynn, P. Pedaksi: Methodology for collecting data about survey non-respondents. *Quality and Quantity* 37(3); 2003: 239–61.

Lynn, P. From standardised to targeted survey procedures for tackling non-response and attrition, *Survey Research Methods* 11(1); 2017: 93–103.

Mavletova, A., I. Deviatko, and N. Maloshonok. Invitation design elements in web surveys: Can one ignore interactions? *Bulletin of Sociological Methodology/Bulletin de Mèthodologie Sociologique* 123(1); 2014: 68–79.

McGonagle, K., M. Couper, and R. F. Schoeni. Keeping track of panel members: An experimental test of a between-wave contact strategy. *Journal of Official Statistics* 27(2); 2011: 319–38.

McGonagle, K. A., R. F. Schoeni, and M. P. Couper. The effects of a between-wave incentive experiment on contact update and production outcomes. *Journal of Official Statistics* 29(2); 2013: 1–17.

McMorris, B. J., R. S. Petrie, R. F. Catalano, C. B. Fleming, K. P. Haggerty, and R. D. Abbott. Use of web and in-person survey modes to gather data from young adults on sex and drug use: An evaluation of cost, time, and survey error based on a randomized mixed-mode design. *Evaluation Review* 33(2); 2009: 138–58.

Mercer, A., A. Caporaso, D. Cantor, and R. Townsend. How much gets you how much?: Monetary incentives and response rates in household surveys. *Public Opinion Quarterly* 79(1); 2015: 105–29.

Millar, M. M. and D. A. Dillman. Improving response to web and mixed-mode surveys. *Public Opinion Quarterly* 75(2); 2011: 249–69.

Montaquila, J. M., J. Michael Brick, D. Williams, K. Kim, and D. Han. A study of two-phase mail survey data collection methods. *Journal of Survey Statistics and Methodology* 1(1); 2013: 66–87.

Munoz-Leiva, F., J. Sanchez-Fernandez, F. Montoro-Rios, and J. A. Ibanez-Zapata. Improving the response rate and quality in web-based surveys through the personalization and frequency of reminder mailings. *Quality & Quantity* 44(5); 2010: 1037–52.

Murphy, S. A. Optimal dynamic treatment regimes. *Journal of the Royal Statistical Society: Series B (Statistical Methodology)* 65(2); 2003: 331–55. doi: 10.1111/1467-9868.00389.

Murphy, S. A., K. G. Lynch, D. Oslin, J. R. McKay, and T. TenHave. Developing adaptive treatment strategies in substance abuse research. *Drug and Alcohol Dependence* 88(Supplement 2); 2007: S24–30. doi: 10.1016/j.drugalcdep.2006.09.008.

Olson, K. Survey participation, nonresponse bias, measurement error bias, and total bias. *Public Opinion Quarterly* 70(5); 2006: 737–58.

Olson, K. and J. Wagner. A feasibility test of using smartphones to collect GPS information in face-to-face surveys. *Survey Research Methods* 9(1); 2015: 1–13.

O'Muircheartaigh, C. and P. Campanelli. A multilevel exploration of the role of interviewers in survey non-response. *Journal of the Royal Statistical Society. Series A (Statistics in Society)* 162(3); 1999: 437–46.

Pasek, J., S. M. Jang, C. L. Cobb, J. M. Dennis, and C. Disogra. Can marketing data aid survey research? Examining accuracy and completeness in consumer-file data. *Public Opinion Quarterly* 78(4); 2014: 889–916.

Peytchev, A. Models and interventions in adaptive and responsive survey designs, DC-AAPOR Panel on Adaptive Survey Design. Washington, DC., February, 2014. Accessed March 26. Available at http://dc-aapor.org/ModelsInterventionsPeytchev.pdf.

Peytchev, A. Responsive design in telephone survey data collection. *International Workshop on Household Survey Nonresponse*, Nürnberg, Germany, August 30, 2010.

Peytchev, A. Interactive case management. *Workshop on Advances in Adaptive and Responsive Designs*, Heerlen, The Netherlands, 2013.

Peytchev, A. and B. Neely. RDD telephone surveys: Toward a single-frame cell-phone design. *Public Opinion Quarterly* 77(1); 2013: 283–304. doi: 10.1093/poq/nft003.

Peytchev, A., R. K. Baxter, and L. R. Carley-Baxter. Not all survey effort is equal: Reduction of nonresponse bias and nonresponse error. *Public Opinion Quarterly* 73(4); 2009: 785–806.

Peytchev, A., S. Riley, J. Rosen, J. Murphy, and M. Lindblad. Reduction of nonresponse bias in surveys through case prioritization. *Survey Research Methods* 4(1); 2010: 21–9.

Peytcheva, E., A. Peytchev, and M. Jans. Collecting proxy measures of key survey variables to estimate, reduce, and adjust for nonresponse bias. Joint Statistical Meetings of the American Statistical Association, Chicago, IL, 2016.

Pickery, J. and G. Loosveldt. A multilevel multinomial analysis of interviewer effects on various components of unit nonresponse. *Quality and Quantity* 36(4); 2002: 427–37.

Pratt, D. J., M. Cominole, A. Peytchev, J. Rosen, B. Shepherd, P. Siegel, D. Wilson, and J. Wine. Using predicted response propensities for bias reduction. American Association for Public Opinion Research Annual Conference, Boston, MA, 2013.

Purdon, S., P. Campanelli, and P. Sturgis. Interviewers calling strategies on face- to-face interview surveys. *Journal of Official Statistics* 15(2); 1999: 199–216.

Raghunathan, T. E. and J. E. Grizzle. A split questionnaire survey design. *Journal of the American Statistical Association* 90(429); 1995: 54–63.

Rao, R. S., M. E. Glickman, and R. J. Glynn. Stopping rules for surveys with multiple waves of nonrespondent follow-up. *Statistics in Medicine* 27(12); 2008: 2196–213.

Roberts, C., C. Vandenplas, and M. Ernst Stähli. Evaluating the impact of response enhancement methods on the risk of nonresponse bias and survey costs. *Survey Research Methods* 8(2); 2014: 67–80.

Romanov, D. and M. Nir. Get it or drop it? Cost-benefit analysis of attempts to interview in household surveys. *Journal of Official Statistics* 26(1); 2010: 165–91.

Rookey, B. D., L. Le, M. Littlejohn, and D. A. Dillman. Understanding the resilience of mail-back survey methods: An analysis of 20 years of change in response rates to national park surveys. *Social Science Research* 41(6); 2012: 1404–14.

Rosen, J. A., J. Murphy, A. Peytchev, T. Holder, J. Dever, D. Herget, and D. Pratt. Prioritizing low propensity sample members in a survey: Implications for nonresponse bias. *Survey Practice* 7(1); 2014.

Rubin, D. B. *Multiple Imputation for Nonresponse in Surveys*. New York, NY: John Wiley & Sons, 1987.

Sanders, H. L., J. Wagner, J. McCarthy, J. Qi, and F. Kreuter. Reducing bias and sampling error: Using simulation to identify effective adaptive design strategies for the crops agricultural production survey. *Paper presented at the TSE15*, Baltimore, MD, 2015.

Särndal, C. E. The 2010 Morris Hansen lecture: Dealing with survey nonresponse in data collection, in estimation. *Journal of Official Statistics* 27(1); 2011: 1–21.

Särndal, C.-E., K. Lumiste, and I. Traat. Reducing the response imbalance: Is the accuracy of the survey estimates improved? *Survey Methodology* 42(2); 2016: 219–38.

Särndal, C. E. and S. Lundström. *Estimation in Surveys with Nonresponse*. Chichester: Wiley, 2005.

Särndal, C. E. and P. Lundquist. Accuracy in estimation with nonresponse: A function of degree of imbalance and degree of explanation. *Journal of Survey Statistics and Methodology* 2; 2014a: 361–87.

Särndal, C.-E. and P. Lundquist. Balancing the response and adjusting estimates for nonresponse bias: Complementary activities. *Journal de la Sociéte Française de Statistique* 155(4); 2014b: 28–50.

Sauermann, H. and M. Roach. Increasing web survey response rates in innovation research: An experimental study of static and dynamic contact design features. *Research Policy* 42(1); 2013: 273–86.

Schafer, J. L. and J. Kang. Average causal effects from nonrandomized studies: A practical guide and simulated example. *Psychological Methods* 13(4); 2008: 279–313.

Schouten, B. Statistical inference based on randomly generated auxiliary variables. Discussion Paper 201515, CBS, Den Haag, 2015. Available at www.cbs.nl

Schouten, B., M. Calinescu, and A. Luiten. Optimizing quality of response through adaptive survey designs. *Survey Methodology* 39(1); 2013: 29–58.

Schouten, B. and F. Cobben. R-Indexes for the Comparison of Different Fieldwork Strategies and Data Collection Modes. Discussion Paper 07002, 2007.

Schouten, B., F. Cobben, and J. Bethlehem. Indicators for the representativeness of survey response. *Survey Methodology* 35(1); 2009: 101–14.

Schouten, B., F. Cobben, P. Lundquist, and J. Wagner. Does balancing survey response reduce nonresponse bias? *Journal of the Royal Statistical Society, Series A* 179(3); 2016: 727–748.

Schouten, B. and N. Shlomo. Selecting adaptive survey design strata with partial R-indicators. *International Statistical Review* 85(1); 2017: 143–63.

Schouten, B., N. Shlomo, and C. Skinner. Indicators for monitoring and improving representativeness of response. *Journal of Official Statistics* 27(2); 2011: 231–53.

Seaman, S., J. Galati, D. Jackson, and J. Carlin. What is meant by "Missing-at-Random?", *Statistical Science*, 28(2); 2013: 29–58.

Shlomo, N., C. Skinner, and B. Schouten. Estimation of an indicator of the representativeness of survey response. *Journal of Statistical Planning and Inference* 142; 2012: 201–11.

Singer, E. and C. Ye. The use and effects of incentives in surveys. *The ANNALS of the American Academy of Political and Social Science* 645(1); 2013: 112–41.

Sinibaldi, J., M. Trappmann, and F. Kreuter. Which is the better investment for nonresponse adjustment: Purchasing commercial auxiliary data or collecting interviewer observations? *Public Opinion Quarterly* 78(2); 2014: 440–73.

Smyth, J. D., D. A. Dillman, L. M. Christian, and A. C. O'Neill. Using the internet to survey small towns and communities: Limitations and possibilities in the early 21st century. *American Behavioral Scientist* 53(9); 2010: 1423–48.

Steele, F. and G. B. Durrant. Alternative approaches to multilevel modelling of survey non-contact and refusal. *International Statistical Review* 79(1); 2011: 70–91. doi: 10.1111/j.1751-5823.2011.00133.x.

Stussman, B., J. Dahlhamer, and C. Simile. *The Effect of Interviewer Strategies on Contact and Cooperation Rates in The National Health Interview Survey*. Washington, DC: Federal Committee on Statistical Methodology, 2005.

Sudman, S. Probability sampling with quotas. *Journal of the American Statistical Association* 61; 1966: 749–71.

Sutton, R. S. and A. G. Barto. *Reinforcement Learning: An Introduction, Adaptive Computation and Machine Learning*. Cambridge, MA: MIT Press, 1998.

Thall, P. F. and S. J. Lee. Practical model-based dose-finding in phase I clinical trials: Methods based on toxicity. *International Journal of Gynecological Cancer* 13(3); 2003: 251–61.

Thall, P. F. and K. E. Russell. A strategy for dose-finding and safety monitoring based on efficacy and adverse outcomes in phase I/II clinical trials. *Biometrics* 54(1); 1998: 251–64.

Tourangeau, R. Mixing modes: Tradeoffs among coverage, nonresponse, and measurement error. In *Total Survey Error in Practice*, edited by P. P. Biemer, E. de Leeuw, S. Eckman, B. Edwards, F. Kreuter, L. E. Lyberg, N. C. Tucker, and T. Brady West (pp. 115–132). New York: Wiley, 2017.

Tourangeau, R., M. Brick, S. Lohr, and J. Li. Adaptive and responsive survey designs: A review and assessment. *Journal of the Royal Statistical Society A* 180(1); 2017: 203–223.

Tourangeau, R., L. J. Rips, and K. Rasinski. *The Psychology of Survey Response*. Cambridge: Cambridge University Press, 2000.

Tourangeau, R. and T. W. Smith: Asking sensitive questions—The impact of data collection mode, question format, and question context. *Public Opinion Quarterly* 60(2); 1996: 275–304.

Tourangeau, R. and T. Yan. Sensitive questions in surveys. *Psychological Bulletin* 133(5); 2007: 859.

Trussell, N. and P. J. Lavrakas. The influence of incremental increases in token cash incentives on mail survey response: Is there an optimal amount? *Public Opinion Quarterly* 68(3); 2004: 349–67.

Valliant, R., J. A. Dever, and F. Kreuter. *Practical Tools for Designing and Weighting Survey Samples*. New York, NY: Springer, Statistics of Social and Behavioral Sciences, 2013.

Vansteelandt, K. and Vermeulen, S. Bias-reduced doubly robust estimation. *Journal of the American Statistical Association* 110; 2015: 1024–36.

Van Vaerenbergh, Y. and Thomas, T. D. Response styles in survey research: A literature review of antecedents, consequences and remedies. *Public Opinion Research* 25(2); 2012: 195–217.

Wagner, J. Adaptive survey design to reduce nonresponse bias, PhD thesis, University of Michigan, Ann Arbor, USA, 2008.

Wagner, J. The fraction of missing information as a tool for monitoring the quality of survey data. *Public Opinion Quarterly* 74(2); 2010: 223–43.

Wagner, J. A comparison of alternative indicators for the risk of nonresponse bias. *Public Opinion Quarterly* 76(3); 2012: 555–75.

Wagner, J. Adaptive contact strategies in telephone and face-to-face surveys. *Survey Research Methods* 7(1); 2013: 45–55.

Wagner, J. Limiting the risk of nonresponse bias by using regression diagnostics as a guide to data collection. *Paper presented at the Joint Statistical Meetings*, Boston, 2014.

Wagner, J., J. Arrieta, H. Guyer, and M. B. Ofstedal. Does sequence matter in multimode surveys: Results from an experiment. *Field Methods* 26(2); 2014a: 141–55.

Wagner, J. and T. E. Raghunathan. A new stopping rule for surveys. *Statistics in Medicine* 29(9); 2010: 1014–24. doi: 10.1002/sim.3834.

Wagner, J., R. Valliant, F. Hubbard, and L. Jiang. Level-of-effort paradata and nonresponse adjustment models for a national face-to-face survey. *Journal of Survey Statistics and Methodology* 2(4); 2014b: 410–32. doi: 10.1093/jssam/smu012.

Wagner, J., B. T. West, N. Kirgis, J. M. Lepkowski, W. G. Axinn, and S. K. Ndiaye. Use of paradata in a responsive design framework to manage a field data collection. *Journal of Official Statistics* 28(4); 2012: 477–99.

Wagner, J., B. T. West, H. Guyer, P. Burton, J. Kelley, M. P. Couper, and W. D. Mosher. The effects of a mid-data collection change in financial incentives on total survey error in the national survey of family growth. In: *Total Survey Error in Practice,* edited by P. P. Biemer, E. de Leeuw, S. Eckman, B. Edwards, F. Kreuter, L. E. Lyberg, N. C. Tucker, and T. Brady West. New York, Wiley, 2017.

Walejko, G. and J. Wagner. Challenges to innovation in face-to-face surveys posed by interviewer noncompliance. *Paper presented at the American Association for Public Opinion Research,* Hollywood, FL, 2015.

Webster, C. Hispanic and Anglo interviewer and respondent ethnicity and gender: The impact on survey response quality. *Journal of Marketing Research* 33(1); 1996: 62–72.

Weeks, M. F., R. A. Kulka, and S. A. Pierson. Optimal call scheduling for a telephone survey. *Public Opinion Quarterly* 51(4); 1987: 540–49.

West, B. T. An examination of the quality and utility of interviewer observations in the National Survey of Family Growth. *Journal of the Royal Statistical Society: Series A (Statistics in Society)* 176(1); 2013: 211–25.

West, B. T., J. Wagner, F. Hubbard, and H. Gu. The utility of alternative commercial data sources for survey operations and estimation: Evidence from the National Survey of Family Growth. *Journal of Survey Statistics and Methodology* 3(2); 2015: 240–64.

Wilson, D., B. Shepherd, and J. Wine. *The Use of a Calibration Sample in a Responsive Survey Design. American Association for Public Opinion Research.* Hollywood, FL, May 15, 2015.

Wolter, K. *Introduction to Variance Estimation.* New York: Springer, Statistics for Social and Behavioral Sciences, 2007.

Index

Y

Z

For Product Safety Concerns and Information please contact our EU representative GPSR@taylorandfrancis.com
Taylor & Francis Verlag GmbH, Kaufingerstraße 24, 80331 München, Germany

www.ingramcontent.com/pod-product-compliance
Ingram Content Group UK Ltd.
Pitfield, Milton Keynes, MK11 3LW, UK
UKHW022223190926
475702UK00003B/486
* 9 7 8 1 4 9 8 7 6 7 8 7 3 *